STORMS, FLOODS
and SUNSHINE

Photo by Whitesell.

STORMS, FLOODS and SUNSHINE

ISAAC MONROE CLINE

AN AUTOBIOGRAPHY

With a Summary of Tropical Hurricanes

PELICAN PUBLISHING COMPANY
Gretna 2000

To
ALLIE MAY, ROSEMARY,
and
ESTHER BALLEW

Copyright © 1945, 1947, 1951
By Isaac Monroe Cline
Copyright © 2000
By Pelican Publishing Company, Inc.
All rights reserved

First edition, 1945
Second edition, 1947
Third edition, 1951
Second printing, 2000

ISBN: 1-56554-766-7

Printed in the United States of America
Published by Pelican Publishing Company, Inc.
1000 Burmaster Street, Gretna, Louisiana 70053

STORMS, FLOODS AND SUNSHINE
Part I
A BOOK OF MEMOIRS

Contents

	Foreword by Stuart O. Landry...............	xi
	Foreword by Nash Roberts	xv
	Preface	xvii
I.	Introduction	3
II.	Ancestry: Dutch, English, Irish, Scotch......	9
III.	Bat Creek Neighborhood....................	12
IV.	Short Bark Neighborhood...................	20
V.	A Student at Hiwassee College..............	25
VI.	Student at Fort Myer, Virginia..............	30
VII.	Assistant Weather Observer and Medical Student	38
VIII.	In Charge of the Weather Observation Station at Fort Concho..................	49
IX.	In Charge of the Weather Observation Station at Abilene......................	54
X.	In Charge of Station, Section Director at Galveston	64
XI.	Summer Hot Winds on the Great Plains......	70
XII.	My Transfer to the Department of Agriculture	74
XIII.	Cooperation of Mexican Weather Service with U. S. Weather Bureau...................	79
XIV.	Temperature Forecasts—The Blizzard of February, 1899	85
XV.	Floods in the Colorado and Brazos Rivers....	89
XVI.	Cyclone of September 5-10, 1900—The Galveston Hurricane	92

STORMS, FLOODS AND SUNSHINE
CONTENTS CONTINUED
PART I.

XVII.	Change from Medicine to Cyclones	103
XVIII.	Transfer to New Orleans	105
XIX.	Floods in the Mississippi River	109
XX.	The Mississippi River Flood of 1903	115
XXI.	Ordered to Honolulu but Orders Revoked	121
XXII.	Art Collector	125
XXIII.	Care of Oil Paintings	130
XXIV.	Oriental Bronze Collection	133
XXV.	Another Mississippi River Flood	136
XXVI.	Political Aspirations of Chief Moore	140
XXVII.	The Explosion	144
XXVIII.	Hurricane Listening Posts	147
XXIX.	Professor Marvin Appointed Chief of the Weather Bureau	153
XXX.	The 1915 Tropical Cyclone at New Orleans	157
XXXI.	Climatic Changes at New Orleans	162
XXXII.	Portraits of Americans by American Painters	167
XXXIII.	The Corpus Christi Hurricane of September, 1919	171
XXXIV.	Tropical Cyclone, June 21-22, 1921, Texas Gulf Coast	175
XXXV.	Studies of Destructive Tides Caused by Tropical Cyclones	178
XXXVI.	My Book "Tropical Cyclones"	182
XXXVII.	Wireless Telegraphy—Radio	189
XXXVIII.	American Portraits—Second and Third Collections	193

STORMS, FLOODS AND SUNSHINE

CONTENTS CONTINUED

PART I.

XXXIX.	The Great Mississippi River Flood of 1927....	197
XL.	Appreciated Comment by the New Orleans Item-Tribune	212
XLI.	An Incident of the Great Flood.............	215
XLII.	Comments on Tropical Cyclone of September, 1938	219
XLIII.	Tropical Cyclone, North Atlantic, 1944........	223
XLIV.	The Garden of Eden and the Ark in America..	226
XLV.	A Voudou Queen...........................	233
XLVI.	Observations on World War II...............	237
XLVII.	Reflections	247

List of Illustrations

Portrait of Dr. Cline........................Frontispiece	
Bronze Bust of Dr. Cline in the Isaac Delgado Museum of Art	4
A Tornado	63
Restoring Portrait of Marie Laveaux ....Opposite	112
Increase in Flood Heights as Levees Are Raised..........	114
Sandbag Levee in Front of New Orleans. Built as the Result of Flood Warnings ....Opposite	128
Mississippi River Crevasse .... "	160
Flood Crest Transferred from River Channel to Overland Flood ....Opposite	192
Bronze Tablet Presented to Dr. Cline by the Southern Pacific Lines ....Opposite	224

(vii)

STORMS, FLOODS AND SUNSHINE
PART II.
Characteristics of Tropical Cyclones
CONTENTS

Introduction: Where tropical cyclones occur and their provincial names 253
Uniform nomenclature for different parts of cyclone...... 258
Influence of topography limits regions where studies can be made 260
Storm tides call for study of tropical cyclones............ 260
Methods followed heretofore do not give correct results.... 262
Integration method used for first time in study of cyclones. 263
Personal experiences in tropical cyclones................ 264
Barometric pressure in tropical cyclones................ 264
Tropical cyclone, August 13-17, 1915................... 266
Powerful water currents developed in the right half of the cyclone 266
Tropical cyclone, September 29-30, 1915................ 267
Tropical cyclone, September 9-15, 1919................. 268
Powerful currents developed in the Gulf in right half of cyclone 272
No records found that report such currents in tropical cyclones 273
Prevailing winds of region carry cyclone forward........ 274
Wind directions as related to the center of tropical cyclones 278
Turbine principle in tropical cyclones.................. 281
Cloud movements in tropical cyclones.................. 283
Cyclones that stopped traveling....................... 284

STORMS, FLOODS AND SUNSHINE
CONTENTS CONTINUED
PART II.

Wind velocities in right and left halves of tropical cyclones 284

Wind velocities in rear of calm center in traveling tropical cyclones 285

Causes and distribution of precipitation in tropical cyclones 289

Influence of elevated land areas, mountains, etc........... 293

The same cause will always produce the same effect...... 296

Origin and development of tropical cyclones.............. 296

Continuous redevelopment of cyclone over area of greatest precipitation 300

Storm tides and currents in tropical cyclones............. 305

Gulf cyclone, September 4, 1949........................ 309

Problem of so-called hurricane busters.................. 311

Radar? Rainfall not around the center of the cyclone...... 312

CHARACTERISTICS OF TROPICAL CYCLONES

Illustrations

Figure 1 Regions where tropical cyclones occur........ 255

Figure 2 Design of gas and whistling buoy moved by water currents developed by winds in right half of cyclone 259

Figure 3 Depth of storm tide extending 100 miles inland 261

Figure 4 Barometric gradients in cyclone do not show decrease in intensity after having moved 100 miles from the coast...................... 269

(ix)

STORMS, FLOODS AND SUNSHINE
ILLUSTRATIONS CONTINUED
PART II.

Figure 5   Height of storm tide along Gulf coast caused by tropical cyclone, September 11-14, 1915...... 271

Figure 6   Ground plan of tropical cyclone from recorded observations 279

Figure 7   Air stream in right rear quadrant of tropical cyclone illustrating the turbine principle in the cyclone 282

Figure 8   Precipitation during the passage of the tropical cyclone, August 15-22, 1915, across the United States 295

Figure 9   Relative sizes and direction of travel of waves and swells developed by winds in tropical cyclones 308

Appendix A   A Century of Progress in the Study of Cyclones 318

Bibliography   Contributors to the study of cyclones...... 337

Appendix B   Editorials 344

Appendix C   Japan's Ally: The typhoon caught the United States Navy unprepared or off its guard. Who was to blame for damage caused by the typhoon? 349

Foreword

FROM plow boy to principal meteorologist of the U. S. Weather Bureau, with medicine and art on the side, sums up the life work of Dr. Isaac Monroe Cline, one of New Orleans' truly remarkable men.

If you stroll down the French Quarter and wander along Saint Peter Street, only a few hundred feet from the old Cabildo you will find an old shop filled with glass bottles, sandwich glass and other items that collectors of glassware admire. Introducing yourself to a pleasant-faced, bright-eyed, old gentleman, you meet Dr. Isaac Monroe Cline. He might even be dressed in his overalls as he works in the back of his shop, restoring an old painting. In any event, he will take time to chat with you about the problems of today or reminisce about the times of long ago. And Dr. Cline is a raconteur of no mean ability. It is hard to realize while you are talking to the dean of the French Quarter that he is eighty-four years *young*. His good humor, his zest for living, and his sane outlook on life are indeed an inspiration to the present-day generation.

Dr. Cline's autobiography tells in an interesting way of the development of the Weather Bureau Service in the United States. For 53½ years he was connected with that service.

Beginning under the auspices of the Signal Corps of the Army, the forecasting of weather became of such importance that a separate bureau was organized and placed under the Department of Agriculture. Dr. Cline was with the service almost from the beginning. He saw it develop from a haphazard guessing of weather conditions to scientific forecasting in which the forecasts are 95% accurate. In addition to the forecasting of weather for the benefit of farmers and business men, and the warning of the approach of freezes so that crops might be protected, part of the weather bureau's service is the issuance of flood warnings. This is particularly important with regard to the Mississippi River floods. The approach of tropical hurricanes and the issuance of storm warnings is likewise an important activity of the bureau.

Dr. Cline, who for so long was in charge of the weather bureau both at Galveston and New Orleans, studied his business and became an authority on storms. He may be called the "cyclone detective". Dr. Cline tracked down and studied thousands of records of wind velocities, of storm tides and other data, and from these he learned the secrets of the cyclone. Bret Harte in 1870 wrote:

> *Said the black-browed hurricane*
> *Brooding down the Spanish Main,*
> *"Shall I see my forces, zounds!*
> *Measured in square inches pounds?*
> *With detectives at my back*
> *When I double on my track!*
> *All my secret paths made clear!*
> *Published to a hemisphere?*
> *Shall I? BLOW me, if I do!"*

Dr. Cline nearly sixty years later fulfilled this prophecy. His famous book, TROPICAL CYCLONES, is used throughout the world as a textbook, and his theory of the turbine principle of tropical cyclones is a contribution to the science of meteorology.

We know, of course, that a man of parts is good in several lines of endeavor. Napoleon was not only a general but he was a good law-maker, a good organizer, a financier, a diplomat and engineer. Leonardo da Vinci was a musician, poet, scientist and painter. He played the flute and wrote poems to please and fascinate the ladies of the court. He constructed military engines to help Lorenzo win victories. He studied the flights of birds, and in his spare time he painted pictures that are among the finest in the world.

Dr. Cline, of course, has been more modest in his achievements, but as a man of parts he is not only a scientist and one of the finest meteorologists in the world, but he also studied medicine and lectured in medical schools. He is a lecturer and author of several books and pamphlets and has written many scientific papers.

In his spare time Dr. Cline became an art collector. As a result the Isaac Delgado Museum of Art in New Orleans contains a fine collection of Oriental bronzes, and the Mellon Collection, which now hangs in the National Gallery of Art in Washington, contains some of the finest portraits painted by American artists. These were rescued from oblivion, preserved from destruction, and were first owned by Dr. Cline before they were sold to Thomas B. Clarke. They later became part of the Mellon Collection. And only recently a beautiful book—THE MASTERPIECES OF PAINTING FROM THE NATIONAL GALLERY OF ART, edited by Huntington Cairns and John Walker, reproduced a Stuart portrait discovered by Dr. Cline. This is a portrait of Mrs. Yates, wife of a New York merchant—1793. The caption says: "One of the finest of Gilbert Stuart's portraits. Collections of Dr. Isaac M. Cline, Thomas B. Clarke, Mellon Collection, 1940".

Thus Dr. Cline deserves the thanks of all American art lovers for his recognition of the importance of early American painters, and his collecting and preserving for posterity many now famous pictures. In addition to the paintings collected and restored, he has helped to restore many fine paintings belonging to private collectors. We marvel at the country boy from the backwoods of Tennessee, not only that he became one of America's leading meteorologists, but that he should have achieved considerable distinction in an entirely unrelated field—that of collecting and preserving masterpieces of portraiture.

In these pages are recorded an eye-witness description of one of America's greatest tragedies—the Galveston hurricane of 1900. The author has also told the story of many of the happenings in the U. S. Weather Service during its first sixty years, and written about his experiences in the great storms and destructive floods that he has witnessed. There are many interesting anecdotes of American life, during an age and generation long since gone. The book contains much historical and scientific matter, as well as many human interest stories.

It is interesting to know that Dr. Cline typed his own memoirs. In fact he retyped parts of his book as many as three times. Think of a man eighty-three year old not only composing his manuscript, but typing it himself! And the final manuscript was neatly done and generally free from errors.

All biographies, no matter how dull or narrow the life of the narrator, contain something of interest to the reader. But the biography of an unusual personality is a record that is perennially refreshing and inspiring.

STORMS, FLOODS AND SUNSHINE likewise points a moral, and that is the importance of the individual in his conflict with his environment. Although the most powerful forces of nature are wrapped up in a tropical cyclone, yet the doughty master of a frail craft will often ride safely through such a storm. And in the storms and battles of life, a courageous soul creates his own destiny and masters the forces that might otherwise wreck or destroy him. Dr. Isaac Monroe Cline is a self-made man. His ambition, his determination, his courage are of the stuff of our American pioneers. His biography is a record of a man who picked out a goal and then achieved it.

—STUART O. LANDRY.

Note

Dr. Isaac Monroe Cline was a member of the following scientific associations:

- Member, International Meteorological Congress, held at Chicago, Ill., August 21-24, 1893.
- Member, Second Pan-American Scientific Congress, held at Washington, D. C., December 27, 1915 - January 8, 1916.
- Fellow, American Association for the Advancement of Science.
- Fellow, American Meteorological Society, President, 1934-1935.
- Fellow, American Geographical Society.
- Member, American Geophysical Union.
- Honorary Member, New Orleans Academy of Science, President, 1935-1936.
- Member, The Astronomical Society of New Orleans.
- Instructor in Medical Climatology, Medical Department, University of Texas, 1896-1901.
- Former Member, Texas Medical Association and American Medical Association.

Foreword

THE author, Isaac Monroe Cline, was born in his family's log cabin farmhouse near Madisonville, Tennessee. His life story could have been written by Horatio Alger. He was a boy born on a small farm shortly before the beginning of the Civil War. He lived in a small area that remained loyal to the North while located in a state loyal to the Confederacy. In spite of the tumultuous conditions that existed during the first five years of his life, he and seven siblings seemed to have had a reasonably stable childhood. At seven years of age, he was required to devote most of his time to working on the family farm. His studies were relegated to a part-time schedule.

Eventually, he completed his elementary education, and at sixteen years of age was admitted to Hiwassee College, located only five miles from his home. This was a liberal arts school that emphasized Latin, Greek, and English. He earned a bachelor of arts degree as well as a masters of arts degree in only five years. To pay for his tuition, he worked as a librarian and assistant janitor. As assistant janitor, he chopped wood for the fireplace and did other chores.

Upon graduation in 1882 he was accepted for training by the U.S. Signal Corps to become a weather observer. The training center was located at Fort Myer, Virginia, just outside of Washington, D.C. Upon completion of this training, he was assigned to Little Rock, Arkansas. His duties were light enough that he was able to enroll at the medical school of the University of Arkansas. He earned his M.D. in 1885, after only two and a half years of study. He completed assignments at two frontier weather stations in west Texas before being transferred to a "First Order Station" in Galveston, Texas. As executive officer of the station, he had four assistants, as well as considerable leisure time, which allowed him to become a part-time instructor in "Medical Climatology" at the medical school in Galveston.

About this time, the responsibility for weather observations and forecasting was removed from the Signal Corps and became the responsibility of the newly formed "Weather Bureau," in the Department of Agriculture. This move had a profound effect upon Dr. Cline's future. All policy-making and administration was lodged within a Washington, D.C. bureau. The Weather Bureau issued orders that all forecasts had to originate from the Washington headquarters.

The weathermen of that day did not fully understand the structure of tropical cyclones, nor did they have an understanding of the dynamics involved in cyclone generation and intensification. No great technical knowledge existed relative to steering mechanisms. An example of this dearth of scientific knowledge was demonstrated in the Galveston hurricane of 1900. The Weather Bureau in Washington, D.C. believed the hurricane would cross Cuba, then head north and later recurve to the northeast, into the Atlantic Ocean. This is known as a "normal recurve." However, this move north cannot occur if a ridge of high pressure lies along an east-to-west axis north of the storm. Washington continued to send out advisories moving the hurricane northeast, when in truth the blocking high was guiding the storm west across the Gulf to Galveston.

As tragic as the 1900 Galveston hurricane was (Dr. Cline's wife died as a result), the hurricane did have a beneficial effect in that it caused Isaac Cline to embark on an intensive study of tropical cyclones.

In 1901 Dr. Cline was transferred to New Orleans, Louisiana, as district supervisor. He then began one of the earliest and most intensive investigations of hurricanes. From this research, he authored a landmark book called *Tropical Cyclones*. The book analyzed twenty-five years of tropical storms and hurricanes that occurred between 1900 and 1924. In addition to *Storms, Floods and Sunshine,* Dr. Cline was the author of many scientific papers.

As a native New Orleanian, Dr. Cline was my local weatherman. He served our area from 1901 through 1935. He was "Mr. Weather" in our uptown neighborhood. Shortly after he retired, World War II broke out, and I went away as a naval officer. After the war, I visited Dr. Cline once or twice at his "Art & Crystal" shop in the French Quarter. His interest was then focused upon painting, art restoration, and antiques.

Dr. Cline was an icon in New Orleans, and the subject of much folklore. He frequently walked from his home to the Weather Office, fielding questions from neighbors all the way. The citizens watched closely to see if he carried an umbrella. A Weather Office joke quoted him as author of the adage, "Don't put rain into the forecast until it starts and don't take it out until it stops."

I feel that Dr. Cline made a substantial contribution to the field of meteorology as a whole, and to tropical storm knowledge in particular. I regret that he did not have an opportunity to use the rare combination of both a medical degree and a meteorological degree to research the relationship between weather and human well-being. He was eminently qualified to explore this vital and still under-investigated field of bio-meteorology.

Nash C. Roberts, Jr.

PREFACE

HAVING long since passed the age when Benuvento Cellini said that every man ought to write his autobiography, I am at this late date constrained to set down in writing some of my experiences, and to express in words some of the philosophy that has guided me through life. In a life filled with action, covering the period of this country's greatest growth and development, I feel in all modesty that my memoirs will be of interest to the general reader.

The keynote of my life has been the utilization of time—particularly the efficent use of "recreation time". And this is the theme of my book.

I have always kept active and busy. When seven years old, I was taken from school to become a plow boy on my father's farm. For seven years I worked long hours in the day, some of it being spent in the country school, with much time at night devoted to reading. When I was fourteen years old, I was sent away from home to attend a college preparatory school. During my boyhood I did not spend much time in play, but my recreation was spent reading useful books. In this way I laid the foundation for future successes and achievement. As a boy I resolved that some day I would write a scientific treatise on some subject—at that time I did not know on what subject—that would be of international interest and importance. This boyhood dream was realized many years later when my book, TROPICAL CYCLONES, was published.

While serving as assistant observer of the United States Weather Service at Little Rock, Arkansas, I spent my off-duty hours, my holidays and vacations in studying medicine at the Medical Department of the University of Arkansas, from which I received the degree of Doctor of Medicine. My idea at that time was to make a special study of the effects of weather and climate on the human organism. This was sixty years ago, and I was pioneering in a field in which investigation has made little progress.

After my graduation in medicine, I was placed in charge of a weather observation station in western Texas, on the fringe of the great American desert. I used my recreation time there in editing and publishing a daily and a weekly newspaper. Later I was placed in charge of the Galveston Weather Observation Station, and while there continued my studies on climate and its effect on human beings. I was a lecturer on medical Climatology at the University of Texas School of Medicine.

The Galveston hurricane of September 8, 1900, changed my objective. I gave up my studies of medical climatology and took up that of tropical cyclones. At that time definite information concerning the action of these death-dealing storms was indeed limited. My work in the Galveston hurricane brought about the establishment of the New Orleans Forecast District, and I was transferred to New Orleans in 1901 as District Forecaster for the middle and west Gulf region. Here my recreation time, when not devoted to the study of cyclones, was spent in collecting and restoring paintings, especially early American portraits. Several of the paintings reclaimed are in important museums in the United States.

Since my work for over fifty years has been with the United States Weather Service in connection with forecasting, and since I am giving here some hitherto unpublished matters relating to forecasting storms and floods, I am calling my Book of Memoirs STORMS, FLOODS AND SUNSHINE. As life itself is made up of rain and sunshine, clouds and storms, as well as fair weather and blue skies, it is fitting that my memoirs covering more than three quarters of a century should bear this title.

In these pages are recorded important events and happenings in the United States Weather Service during its first sixty years of existence. There will be found stories of great storms and destructive floods, as well as stormy times in the service. There is much of scientific, historical and human interest.

My autobiography is written at a serious time, when the future of our country, as well as our western civilization is at stake. Looking back over the years, I can recall many moments when it seemed that all was lost. But events finally righted themselves. I have unbounded faith in human nature, and am firm in the belief that some day the world will achieve permanent peace and that the whole human race will enjoy an even higher type of civilization. This will only be brought about, however, by the industry and thrift of individuals, and by the exercise of that intellectual curiosity that makes for progress as well as the genius for invention.

For several years I have had in mind writing these reminiscences. Finally in July, 1943, the plan of the book was laid out so that each of the 47 chapters would be a statement of essential matters relating to some particular incident in my life. The manuscript was practically completed October 13, 1944, just a little more than one week being devoted to writing each chapter. The chapters are somewhat unrelated. This will not detract from the lessons which my story is intended to teach. The reminiscences have been written in a conversational style—in the same style I have used in giving after luncheon talks. This will best present the impressions which I desire to leave with the reader.

I have kept numerous records and notes, and the records in my possession will confirm my statements. In conclusion, I apologize to the general reader for the inclusion of some scientific data, which is of interest to only a limited few. But the study of the weather, of storms and cyclones and floods is so much a part of my life history—it is in fact a part of me— that I had to include some matters that might otherwise never see the light of day.

I. M. C.

STORMS, FLOODS
and SUNSHINE

CHAPTER I.

INTRODUCTION

TENNESSEE, like all Gaul, is divided into three parts. There is eastern Tennessee. This is the Tennessee of the Great Smokies, with its Smoky Mountain ballads, where the people live in log cabins built to keep off the Indians, where they still hold quilting bees, and the mountaineers speak the language and idiom of the time of Queen Elizabeth.

Middle Tennessee is the bluegrass region, with its blooded horses and beautiful farms. The western part of Tennessee is the cotton section, where Memphis, said to be built on the site where DeSoto first saw the Mississippi River, lords it over the middle Mississippi Valley.

Tennessee lies between the 35° and 36° and 30' of north latitude in that part of the north temperate zone where climatic conditions are especially favorable for the physical and mental development of man.

It originally belonged to the Colony of North Carolina, and was part of the grant given to Sir Walter Raleigh in 1584, as well as a later grant made by Charles II in 1663. Although "ancient" territory, so to speak, no permanent settlement was made there until the middle of the Eighteenth Century.

After the Revolution settlers drifted westward over the Great Smoky Mountains. Soon there grew up the populous community known as the Watauga Settlement. Many do not know that this community along with other settlements in 1784 seceded from North Carolina and organized the State of Franklin. This Franklin State existed for four years. Counties were created and taxes were levied. John Sevier was governor until he was arrested in 1788 and allowed to escape. Then the state disintegrated, as it were, and North Carolina resumed its jurisdiction over the State of Franklin. Two years later North Carolina ceded all the territory embracing the State of Tennessee, including the former State of Franklin to the United States. Tennessee was admitted to the Union in

Photo by Whitesell.
Bronze bust deposited by the United States Government in the Isaac Delgado Museum of Art, New Orleans, Louisiana

June, 1796, and was the third state after the original thirteen to join the Union.

Tennessee's population was made up of English, Scotch, Irish, and a few Hollanders who had trickled down from the region about New Amsterdam through Pennsylvania and Virginia. The early settlers were of all trades and professions but mostly farmers. They were a people with the highest standards of honor and honesty. Many religious denominations were represented, some of them worshipped together.

Monroe County, where I was born, was named after President Monroe. Bordering on the Unaka or Great Smoky range of mountains to the east, it is mostly hills and valleys. Its largest range of hills was known in my day as "Bat Creek Knobs" from the fact that they run parallel to Bat Creek, on which was located the Bat Creek settlement.

The valleys along the Little Tennessee River, its tributaries and their feeders, were fertile. The hillsides near the streams where the slopes were not too steep were cultivated. Fork Creek just west of the "Bat Creek Knobs" with much fertile bottom land attracted many settlers, and was soon well populated. Short Bark was another early-formed community. The lay of the country was undulating and the greater part of the region was wooded with a variety of timber. Springs and small streams furnished an abundance of water. The soil was not so productive as that near the larger streams, but with a little fertilization it yielded good crops. It was well suited for growing fruits, such as apples, pears, peaches, plums, etc. It was admirably suited to live stock farming.

My father often related the following story of the old days. In 1820 residents of some of the settlements in Monroe County were wealthy, and imported their clothes from Europe. One of these settlements was Walnut Grove, on the fringe of the Short Bark area. With a good water supply furnished by a small tributary of the Little Tennessee River, Walnut Grove was quite a manufacturing center, with tanneries, machine shops, potteries, etc. This locality was one of the places where the great preacher, Lorenzo Dow, stopped to preach on his journeys back and forth, from the Atlantic Coast to Louisiana, on which he spread the word of the Gospel to the residents of the newly settled regions. Lorenzo Dow traveled by muleback, and a colored servant named Sambo attended him on his journeys to care for the mule and minister to the needs of his master. Sambo was a good trumpeter and on approaching the Reverend Dow's stopping places would sound his trumpet to inform the people that the great preacher would soon arrive. Reverend Dow had misgivings concerning the seriousness of the religious pretences of the congregation at Walnut Grove; he thought that they attended church more to display their fine clothes than to worship the Lord, so he decided to test their faith and acts.

On this particular occasion as he approached Walnut Grove, he instructed Sambo to take his trumpet, and, before daylight Sunday morning to climb a tree in the rear of the church, get into the belfry and remain there until he heard the preacher say, "O Angel Gabriel, sound your trumpet!" Then Sambo was to blow the trumpet with all his might. Reverend Dow with an appropriate text preached a powerful sermon on the second coming of Christ. No other man could have been more inspired. When he had finished his sermon he said: "All who are ready to go to heaven and meet their Lord stand up." Every person in the audience was ready, and they all stood up. He then prayed a fervent prayer, concluding with a loud and emphatic appeal, "O Angel Gabriel, while all these souls are ready to meet their Lord, sound your trumpet!" Sambo made the welkin resound with the music of his trumpet. The congregation stampeded, and not a single person remained in the church to join the Angel Gabriel in the ascent to heaven.

Bat Creek, Fork Creek, and other streams furnished water power for grist mills and saw mills. These mills had large overshot wheels with buckets, which were filled with water from the mill race until the weight of the water accumulated in the buckets started the wheel to turning slowly, and soon the entire machinery of the mill was set in motion. Some of these overshot wheels were 100 feet in diameter. Farmers took their corn and wheat to the mill, the miller took out his toll, and ground the remainder while the farmer waited for his grist.

The mills were simple, but the large wheels through which the water exerted its power, were impressive and had a grandeur about them that was awesome. The introduction of the turbine wheel where the force of the water rather than its weight functioned as the propelling power, did away with these beautiful and stately old wheels.

There were no slaughter houses in those days. Each farmer slaughtered his beef, pork, and mutton, and cured it in his own smoke house. The pioneers knew a great deal about meat preservation, and the meats which they cured were much superior in flavor to the meats turned out by the slaughter houses of today.

There were no regulations to confuse the people in those days. Laws were simple and pertained to the more serious crimes. It was then in reality a country of freedom and free enterprise, where every person understood that his success depended solely upon his own efforts. This condition bred satisfaction and contentment. With all the improvements of today people do not appear as happy and well satisfied as they did seventy-five years ago.

The houses, with a few exceptions, were built with hewn logs, and the cracks were chinked with chips and blocks held in with clay. Lumber was used for flooring and sometimes the more pretentious houses were weather-boarded and ceiled. Neighborhoods which had been long settled and had grown into villages and towns possessed brick buildings.

Steamboats plied the Little Tennessee River, and farmers hauled their produce to Fowler's Landing where the boats picked it up and took it to Knoxville which was a good market. When we had produce at the boat landing we would be on the alert for the sound of the boat whistle. Then we would dress, mount our mules, and ride quickly to the boat landing to check in our shipment. Later the East Tennessee and Virginia Railroad was built through that section, and this gave a more diversified and convenient means of transportation.

The principal towns of Monroe County were Madisonville, the County Seat, Sweetwater and Loudon. There were settlements with stores at Johnson's Mills on Bat Creek, McCroskey's

Mills on upper Fork Creek, Fowler's Mills on lower Fork Creek, Porters on the fringe of Short Bark, and a few others some distance from where my family resided.

As a man is partly the product of his environment, I have tried to give a picture of the land in which I was born and where I spent my early childhood.

I am proud to say that I came from Tennessee, one of the great states of the Union. There Daniel Boone made history as a famous pioneer and Indian fighter. There David Crockett was born. From there came Sam Houston, the founder of Texas. And Tennessee was the home of Andrew Jackson — known as "Old Hickory" — one of America's most forceful and colorful presidents.

Tennessee was young, being only sixty-five years old, when I made my advent into the world on October 13, 1861. As the State of Tennessee is 149 years old at this writing, the span of my life has covered more than half of this time. However, the scenes of my work and my activities have been mostly away from my old home state. But in retrospect I look back into a time long gone, and sigh for the days that are no more.

CHAPTER II.

ANCESTRY: DUTCH, ENGLISH, IRISH, SCOTCH

GENEALOGY is dull reading to all except those interested. The "begot" chapters in the Bible are often skipped. But as a man's ancestry is more important than his environment, it is permissible to include in these personal memoirs a few words about my people.

Several families surnamed Cline were listed in the first United States Census (1790) as residing in that part of Pennsylvania bordering on New Jersey. A few of the Cline families had migrated to Virginia.

My grandfather, John Cline, was of Dutch descent, and his ancestors were among the settlers who came to New Amsterdam as a place of refuge from persecution. His father moved to Virginia after the War of the Revolution.

John Cline married Mary Hawk, a Virginian of English ancestry, about 1826. Mary Hawk was a second cousin of President James Monroe. One of her sons was christened James Monroe. My grandfather was a farmer and owned part of the land on which Bristol, Virginia, now stands. Just after my father, Jacob Leander Cline, was born in 1833, the fourth child of John Cline and Mary Hawk, they moved to Monroe County, Tennessee, and bought a farm on the eastern edge of Short Bark. This was in 1836. John Cline died of smallpox there in 1864. His wife, Mary Hawk Cline, survived him some thirty years and died on the same farm in her early nineties. (The family Bible which contained the complete family history of the Cline and Hawk families was lost soon after my grandmother Cline's death.)

George Wilson, of the South Carolina and Georgia family of that name, married Rebecca Harris in 1834. She was the daughter of a wealthy planter residing near Madisonville, Tennessee, an early settler and one of the few slave owners of that region. My mother was the third child of George Wilson and Rebecca Harris. The Wilson and Harris families were large scale farmers for that time.

Jacob Leander Cline received such education as the schools of that neighborhood of one hundred years ago afforded. He often remarked that his education was limited to the "Three R's—Readin', 'Ritin' and 'Rithmetic". In these elementary subjects he was well grounded. However limited his education might have been, he was a successful man in more ways than one. He worked on his father's farm until he was twenty-one years old. Then his father gave him a horse with a saddle and bridle, and he started out for himself. He was first employed at the Johnson Mills to manage and operate a water power saw mill on Bat Creek about four miles above its junction with the Little Tennessee River.

He was paid $4.00 a week and board for this work. He said he studied at night by the light of a pine knot, then took his books to the mill with him where he fixed a seat to the log carriage so that he could sit there and study while the carriage was feeding the log to the saw, and at the same time be ready to reverse the carriage when the end of the log reached the saw. Young Cline worked four years in the Johnson Mills, during which time he studied the operation of the stationary steam engine, then being used to operate mills in regions where there was no water power. Because of his knowledge of steam engines he secured a position as superintendent and engineer at a much better salary of a saw mill and corn grist mill operated by steam power. This mill was located about ten miles southwest of Madisonville, the County Seat of Monroe County. My father saved his money and bought a small farm near where he worked. He then married Mary Isabel Wilson in 1860. He took her from one of the most imposing homes in that section of the country to the farm where their residence consisted of a log cabin with two rooms and a built-in porch which was used for a kitchen and dining room. Here I was born October 13, 1861.

I was the oldest of eight children: Tiny Alice, Margaret Clementine (died of pneumonia when a child), George Washington, Joseph Leander, Sarah Rebecca, Thomas Alexander, and Cora Della.

Everything was in turmoil and strife as a result of the War Between the States. Son was arrayed against father and brother against brother. My great-grandfather Harris was a slave owner, and cast his lot with the Confederacy, while his sons and many others in that section opposed the dissolution of the Union. This section of Tennessee held elections and was represented in the United States Congress throughout the War Between the States. Lincoln in his declaration of war with the Confederate States specifically excepted that part of Tennessee. Neither the armies of the Confederacy nor of the Union occupied areas nearer than Loudon and Chattanooga, so that there was no destruction of property. Food was scarce and there was considerable suffering, but East Tennessee was not long in getting back to normal conditions.

CHAPTER III.

BAT CREEK NEIGHBORHOOD

FATHER sold his farm soon after the war ended, and rented a place from the Hudson family on upper Fork Creek. After one year he then leased a farm on Bat Creek about two miles above the Johnson Mills where he had worked some ten years earlier. The many hills along this stretch of Bat Creek were referred to as "knobs". Some were so steep that they could not be climbed except by pulling oneself up with the aid of the undergrowth. The knobs were alive with game and wild fowl. The creeks offered excellent fishing, and fur animals, such as muskrat, mink, and otter were plentiful. Quail, some pheasants, and wild turkey afforded sport since unequaled.

Fine drinking water gushed in abundance from springs that issued from under rocky ledges. So great was the flow of water that large streams called "spring branches" were formed. The spring water was cold in summer, and near the air temperature in winter. This water came through limestone rock, and furnished the lime which contributed to the building of large boned, tall men and strong human bodies.

The low lands along the creek were fertile and exceptionally productive. However, damage often resulted from floods, and the crops had to be replanted. Sometimes floods at harvest time ruined matured crops. Hillsides, where not too steep for cultivation, had been protected from erosion and yielded good crops. Father fitted himself out with good equipment and had fine work mules. Land not suitable for cultivation was stocked with cattle and sheep. Two colored families living on the farm were employed by the year. Other men were hired for the plowing and harvesting periods.

Farms in the "knobs" were of small acreage. A certain farmer cultivated one of these knob farms with a pair of oxen. He needed some fire wood but one of his oxen was sick. Deciding that he could take the place of the sick ox, he hitched the well ox on one side of the wagon tongue and himself on the other. Things went nicely while going up the hill, but after the wagon was loaded with wood and the "team" started down the hill, the loaded wagon pressed them forward and the ox started running, carrying the man and wagon down the slope at an uncomfortable speed. The farmer became frightened and yelled at the top of his voice: "We are running away, stop our damn fool selves." A neighbor heard his appeal for help and stopped the ox-man team thus preventing a catastrophe.

Father and mother often told me that I was a problem and hard to handle even before I could walk. My inclination for research evidently developed at an early age; they said I investigated everything that came within my ken. Pottery jugs with stoppers I could not remove were broken to find out what was inside them. When I was about three and a half years old I failed to show up for supper one day. A hearty eater and always on hand for meals, they were convinced something was wrong with me. Mother remembered that I had been telling her about a red apple on one of the trees in the orchard and to the orchard they went in search of the missing boy. They found me unconscious under the apple tree where I had fallen across a log. A limb with the red apple on it was under me, and both my wrists were broken. I got what I went after but paid dearly for the apple.

Bulging Branch school house, built of logs, was about three miles from where we lived. Public schools held short sessions during the winter, and teachers, would come from distances and hold pay schools in the interim. One of these teachers was a Mr. Monroe of Virginia, a cousin of my grandmother Cline. He was a fine looking man with white hair reaching down to his shoulders. Father and mother decided that the only way to keep me under control was to place me in school at the age of four years. They arranged for me to go along with and be cared for by some older children of the Lowry family who lived about a quarter of a mile up the creek. I was large for my age and was very proud of the opportunity to go along with the older children to school. When Mr. Monroe taught school he boarded with us and I went with him. He was fond of me and I liked him. During the noon hour he would take his chair to a shady spot, lean back against the school house and take a nap. I would get a twig from a bush, twist it into his hair and get out of sight before he was awake. He apparently suspected me for he never tried to find the culprit.

I was studious from the start because I wanted to be as smart as the older children. Father would get me out of bed at 4 a. m. so that I could help him feed the live stock and help mother milk the cows. I helped mother milk and assisted father in taking care of the live stock in the evenings as in the mornings. This left me very little time for mischief. However, I had Saturdays and Sundays during which I found considerable time to play. Father worked in the fields until darkness forced him to quit, and it was often 9 or 10 o'clock at night before I got to bed.

Father employed some colored families, who had been the slaves of my great-grandfather Harris, to live on the farm and work for him. The women folk had helped care for my mother when a child, and were fond of her. The colored mammies were always solicitous about the children of the families of their owners, when the owners had been good to their slaves. The colored women were employed to do the heavy work around the house, but mother did her own cooking and she was an expert. The farm hands were given their dinners and they were well fed. Father kept his hired hands busy working from daylight till dark, but the good dinners compensated for the long hours of hard work. Father always worked with them and kept them on the job. One of the colored men was a blacksmith and operated the blacksmith shop. Here I spent some of my spare time. I learned to be a first-class blacksmith, and what I learned there has been valuable to me in many ways all through life. The head of one of the colored families was an old man with grey hair. He was too old to work and spent most of his time fishing and trapping. He sold his pelts for good prices. Father did not want me to fish and trap; he said that men who fished and trapped seldom accomplished much in life, and he wanted me to do something worth while. He impressed me early in life that I should prepare myself so that I would be able "to go places and do things". At that time I was my father's great hope for the next two children were girls.

Notwithstanding father's admonitions against trapping and fishing, the spirit of Isaac Walton surged within me. Saturdays, when home from school, I would steal away and go with the old colored man to see his traps and watch him fish. I had no money with which to buy fish hooks. The blacksmith shop came to my aid. I took large pins, bent them into shape, and filed a beard on them to keep the fish from getting off, and with these crude fish hooks I caught some fish. The old colored man would mix corn meal dough with cotton and throw it into places in the creek where he fished. He told me he was baiting the fish, that they would eat that corn meal and while they were waiting for more he would catch them with his hook. Plausible, yes, and as I was between five and six years old, I decided to try out this idea. There was a nice eddy in the creek about four feet deep with a shoal just below it—a nice place for fish to congregate—about one hundred yards from the house. Here I decided to bait for fish. I took the wash basin filled with dough mixed with cotton, and directly to the creek I went. When I started to throw the dough into the water my foot slipped and the pan with the dough slipped out of my hands into the creek. I shed my clothes and waded in to get the pan, but the water was well up under my arm pits so that I could not reach down to pick up the pan without my head going under the water. I was afraid I would drown if I put my head under the water. I then dressed and went to the house as innocent as a lamb. I did not mention what had happened. Mother called father when he came in that evening and told him the wash basin was missing and that she had not been able to find it. She had not mentioned it to me, therefore, it was evident that she suspected that I had something to do with the disappearance of the pan. Father said: "Son, come with me". He took me directly to the spot where I had dropped the pan in the creek and said, "Son, pull off your clothes and wade in and get that wash basin you dropped in here." I told him that it was deep and I would drown, and thus convicted myself. He made me go in after that pan. The water was well up under my arm pits at the point where I located the pan with my feet, and when I reached for it my head went under the water and I came up frightened and crying, saying that I would be drowned if I went down after that pan. Father said, "Stop that crying and get that pan"! I knew him well enough to realize that he meant what he said, and no fooling. Fortunately, I stepped on the edge of the pan and it turned up which gave me an idea. I held that foot there and slipped the other foot under the pan so that I could raise it high enough to reach it with my hand. We walked home without a word being spoken. When we reached the house and the wash basin was restored to its place, I asked father how he knew that I had dropped the pan in that spot. He said we sometimes know things and cannot explain how we know them. He then told me another story about Lorenzo Dow. On his way to a Lutheran church to preach one Sunday, the Reverend Dow saw a man named Johnson in the woodside hunting for something, and asked him to come to church. Mr. Johnson replied that he had left his iron wedge on a stump nearby and some one had evidently moved it. Since he had promised brother Jones to help him split rails Monday, he would have to locate the iron wedge before he could go to church. Reverend Dow said: "Come and go with me to church and I will help you find your iron wedge." Mr. Johnson went along. On the way to the church the preacher got off his mule at the crossing of a small stream, and picked up a round rock about the size of a goose egg, which he put in his saddle bag. When he took out his Bible and opened it on the pulpit he placed the rock in front of the Bible. He took as his text "Thou shalt not steal", and, being a great orator, he preached an effective and touching sermon. Just as he finished the sermon, he picked up the rock and said: "A man in this church stole Brother Johnson's iron wedge and I am going to hit him with this rock." As he drew back his hand in the apparent act of throwing the rock, a man dodged under one of the benches. The Reverend Dow walked up to him and took the iron wedge out of the man's pocket. Father then said: "You never realize who knows what you are doing, but God always knows and you will be found out." This was probably the best lesson I ever learned.

When about six years old I concluded that I must make some money. Trapping for fur animals appeared about the only thing I could do to get a little cash. Traps were necessary, but I had no money to buy steel traps; so I learned to make the trap known as the "Deadfall", a very effective trap which by its weight mashed the animal flat so that almost instantaneous death resulted—a humane trap. I would steal off Sunday afternoons, locate the "slides" of the animals and fix my traps. I caught muskrat with these traps, but mink and otter were too cunning for that device. I got enough muskrat pelts to buy some steel traps with which I had better luck. I did not dare let this interfere with my routine work. I would get out of bed early enough to visit my traps and get back home by four o'clock in the morning to help feed and care for the livestock and do my other chores. It was exacting but exciting sport. Father permitted me to trap as long as I did not let it interfere with my chores around the house or my school work. I also trapped for game birds, and they were fine eating. From the time I was four years old until I was nearly seven, I was kept in school throughout the year. I studied the three "Rs'" penmanship and drawing and was fast becoming a charcoal artist. This was to be of material service to me in after years.

When I was seven years old, a plow hand whom my father had employed for the crop season, failed to show up for work, so I was taken from school and placed behind the plow. During the plowing season I had been in school and knew nothing about plowing. Father had a mule named "Jim" which he always hitched to his plow. "Jim" was a smart mule; he knew all about plowing, and had more sense than any animal I have ever seen, and more than some people I have known. He knew all the tricks in plowing and took pride in doing good work. This mule taught me how to plow. When I would say, "Whoa, Jim", the mule would stop instantly. At the end of the row of corn, I might give the wrong command, but "Jim" would turn correctly, straddling over the stalks of corn without injuring a blade and then start plowing another furrow. Yes, "Jim" did the plowing and I followed between the plow handles to hold the plow in place. With "Jim's" help I was a first class plow hand for several years. I have often thought that mules belonged to some union. When the noon bell sounded they would double time it to the end of the row and there they would stop. It was almost impossible to make them plow another row until they had their dinner. After I left Tennessee "Jim" was put on the retired list. He was turned loose in a good pasture with plenty to eat. Some years later on a trip back to Tennessee, I went to the pasture to see "Jim". He showed that he recognized me and was glad to see me.

Stories of ghosts were common talk in those days. Yes, I saw a ghost about 2 o'clock one morning when I was about eight years old. There was a plot of timber land at the top of a hill one mile from where we lived, separated by a roadway from the farm my father operated. A short distance from the road, in dense timber, a Mrs. Mowrey had dug for hidden treasure. Rumor was that ghosts had frightened her away before she finished her search. It was said that persons passing there in the night had often seen ghosts. Father called me out of bed about 2 o'clock one morning and told me that I would have to go to Johnson Mills, about two and a half miles distant, to get some turpentine for a mule sick with colic. He put me on the back of the mule, "Jim". I felt safe on Jim's back, for had he not taught me how to plow! But I had to run the gauntlet of the ghosts. There was a steep hill sloping down for nearly half a mile from the place where the ghosts were supposed to make their appearance. As I started down this hill a large white ball, which looked to me as big as a hogshead, started rolling down the hill in front of my mule, keeping about fifty feet in front of us. My mule did not shy, and, although I was frightened, the mule gave me confidence. For half a mile this ghost kept me guessing. When we reached the foot of the hill the white ball rolled into a fence corner and stopped. I screwed my courage up and directed "Jim" directly towards it. As we got closer "Jim" pricked up his ears and my hair stood on end. I said, "Whoa, Jim", patted him on the shoulder, and he went up close to the object. And what do you think it was? Instead of a ghost as large as a hogshead, I saw a bunch of white weeds not larger than a water bucket which the winds had uprooted and carried down the hill in front of us. Since then no ghost has crossed my path, but terrible catastrophies have been encountered as I traveled the journey of life, and I have met stern realities where real danger lurked. This imaginary ghost made me resolve to investigate whatever seeming mystery loomed across my path. I made up my mind that I would never cross bridges in my imagination but to wait until I arrived at the real bridges and then cross them fearlessly. Further along I will describe the crossing of bridges, some fraught with great danger, and one in particular which I crossed after I had given up completely for a moment the idea that it could be done.

CHAPTER IV.

SHORT BARK NEIGHBORHOOD

HARD work and frugal living had enabled father to accumulate some money. He bought two adjoining farms in Short Bark, about two miles from the farm he had been renting, and moved to these farms when I was thirteen years old. One of the farms had a rather pretentious residence situated on top of a small hill, and there were spacious outhouses and barns. About fifty yards from the house, at the foot of the hill, a superb spring of cold water poured forth from a ledge of limestone rocks. Just below the spring was a spring-house especially fitted to keep milk, butter, and other good things cold during the warm months. There were two fine apple orchards, separated by a public road which made them easy to reach. We dried apples in the sun, made cider with a hand press, and other good things to eat and drink. During cider time a large pitcher of cider was placed in the spring-house where it was kept cool and fresh. However, the cider when kept in the spring-house for some time, would commence forming bubbles, and developing alcohol, and it was then at its best. Barrels of boiled cider were prepared in season to be ready for use during the winter.

Father now put into effect his plan to give his boys an education and to give each of his daughters a farm at their marriage. About six miles distant on Fork Creek there was a school where boys and girls went to prepare for college under the direction of Professor Lundy. Arrangements were made to send me there for two years when I would be sixteen and could enter college. Father did not feel financially able to meet the expense necessary to put me in one of the boarding houses. Two other boys of limited means joined me, and we rented a log-cabin. This cabin was situated at the foot of a hill covered with a dense growth of scrub cedar on one side and on the other was a field cultivated in corn and oats in alternate years. The "shack" had two rooms, in one of which we slept; the other room had a rock fire place with a wood and mud chimney, and in this we cooked, ate our meals and studied.

Soon after we rented the house we were told that it was vacant because it was haunted. Ghosts would come out of the cedars and frighten the tenants so that they would leave. One of my uncles had given me an Army Colt six-shooter and plenty of ammunition for use in the case of an attempted holdup, for we were some distance from the school house and no other house was nearer than a mile. We did not give much thought to ghosts, for we did not believe in ghosts. However, one dark night about 10 p. m. rocks commenced raining on the roof of our "shack" attended by moans and screams. There must have been a dozen or more in the crowd from the way the rocks fell on the roof and the noises made. I took my pistol, stooped low and crawled along with the fence between me and the cedars for protection against the falling rocks. The fence furnished protection until I was out of range of the missiles. I located the point in the cedars from which the rocks and screams were coming. I did not want to wound any one and I commenced shooting into the cedars over their heads. With the first shot the rocks and screams stopped and I heard boys running through the cedars. I continued sending bullets through the cedars over their heads and by the time I had fired six shots they were well beyond hearing distance. The ghosts never came back and we never learned their identity. As far as we knew the episode was never mentioned. We kept our part in the affair a secret otherwise my pistol might have been taken away from me.

During the school term I would go home Friday after school and help father on the farm Saturdays. From the time I can remember I was taken to Sunday School and Church regularly, either Cumberland Presbyterian, Lutheran, or Presbyterian. The teachings received in this way had great bearing on my future life. After we returned from church and had a good dinner father would take me back to school and bring with him plenty of good things to eat which mother had fixed up to supplement our cooking during the coming week. I finished the course under Professor Lundy and was ready to enter college.

About this time some one gave me a small black dog named "Rat", with the assurance that he was a good possum dog. Some colored families who lived on the farm would often bring us a possum baked with sweet potatoes, which is a dish fit for a king. "Rat" was a smart little dog, not much larger than a big rat. Every one became much attached to him. During the fall and winter I would take him and go possum hunting Saturday nights. We seldom went hunting that "Rat" did not pick up a trail; when he struck the trail of the possum he would give a sharp bark which he kept up at intervals while he followed the possum to its home in a hollow log or treed it in a persimmon tree where it was feeding on luscious persimmons. If in a log I would chop it out, and when in a tree I would climb the tree and make him drop to the ground. When a possum wrapped his tail around a limb he could not be shaken loose, but when rubbed with a stick on the back of his neck he would let go with his tail and fall to the ground. "Rat" would keep him sullen, playing possum, until I came down out of the tree. On one occasion "Rat" followed a trail something like a mile, and I was sure he was on the track of a Raccoon. When he treed it up a tree nearly two feet in diameter I was certain that it was the Racoon. Some two hours were consumed in chopping the tree down, as it was too large to climb and I did not want to get up in the tree and come in contact with a Raccoon. When the tree fell "Rat" hurriedly went up through the tree limbs and I was behind him with the ax to kill the Raccoon. "Rat" found a hollow in the tree and pulled out a possum about as large as an ordinary house rat. That was the nearest I ever got to catching a raccoon. When we caught a possum we carried it home alive; its tail was put in a split stick and it was held at a distance so that it could not bite us. The possum when killed is placed in boiling water, then quickly removed and the hair scraped off, the same as is done with a hog. It is baked with sweet potatoes and then —um-m!

Mother wove the cloth from cotton and wool grown on the farm and made my clothes until I was fourteen years old. At that age father gave me a plot of ground which I cultivated in wheat or rye on Saturdays and during vacation. I could use the money received from the crops as I wished. With this money I bought my first "store" clothes, and for a country boy I dressed in style.

From the time I could read well, which was early in life, the Bible was my favorite book, and when at home I would study it at night by a pine knot light. The writings of Jules Verne fascinated me and impressed me so strongly that I almost believed in him as a prophet. His imagination thrilled me and inspired me at an early age with the urge to write a book. When working on the farm during school vacation, I carried my books with me and studied while my mule rested. Father would come along and remark: "Son, your mule gets tired very often." In fact I was a dreamer and would plan then what I wanted to do through life. My ambition was to write a book on some scientific subject that would be permanently helpful to man. No definite line of work was in my mind at that time. I was preparing the foundation on which the structure of my life could be built in such a way as future circumstances might determine.

Sunday school and church work had a fascination for me. I was fond of the Bible, and was allowed to interpret it from my own point of view. When I was ready to enter college many of my friends wanted me to become a Cumberland Presbyterian preacher, and endeavored to persuade me to enter one of their schools where preachers were educated. My parents did not try to influence me and I declined because I wanted to be free from any obligation.

Stormy weather was of frequent occurrence in this locality. Forked flashes of lightning and rolling thunder, vibrating through the hills, fascinated me. Father would refer to the thunder as Jupiter Pluvius announcing the approach of the corn wagon, because at the season of the year when thunder storms were frequent rain was generally needed for the corn crop. However, some storms left death and destruction in their wake. One Saturday night a tornado, with its funnel-shaped cloud and destructive swirling winds, moved down the Fork Creek Valley. Many residences were destroyed and several people were killed. The destruction was so great that messengers were sent out over the surrounding country for help. Father went early Sunday morning to render what assistance he could and took me with him. The brick residence of one of father's friends had been demolished and some of the family killed. A child was asleep in a bed on the second floor of the building. The tornado picked up the bed with the child and deposited them, with the bed intact and the child unhurt, in an orchard about one hundred yards from the house.

Mule rustlers were stealing mules in Kentucky and Tennessee and running them down into Georgia where representing themselves as mule dealers, they sold the mules. The theft of mules was so frequent, at this time, that the Tennessee Legislature passed a law which made horse stealing punishable with death on the gallows. The first mule thief caught, was promptly tried, convicted and hanged. That stopped the stealing of both mules and horses. Law enforcement prevents crime, while failure to enforce the law breeds crime. The criminal should always be made to suffer for a crime committed. Prompt and certain punishment deters criminals.

CHAPTER V.

A STUDENT AT HIWASSEE COLLEGE

HIWASSEE College, a school with a national reputation for its thorough teaching of English as well as Greek and Latin, and other subjects which constitute the foundation of learning, was about five miles from where I lived. It was a school where Methodist preachers were educated, but its curriculum enabled young men of limited means to prepare a foundation of learning on which they could build through the future. Its courses in mathematics, the sciences, and languages were good. There were selective studies which led up to civil engineering, meteorology, law, and other fields of activity.

I entered Hiwassee College when I was sixteen years old, and took the liberal arts course leading to the degree of Bachelor of Arts.

In part payment for my tuition, I served as Librarian and assistant janitor, in which latter office I chopped wood for the fire-places, and performed other chores. With another boy I rented a small two-room building on the campus in which we studied, slept, cooked and ate our meals. There were several of these "hutments" we would call them today, which the college rented to poor boys who were striving to get an education.

Mathematics, physics, chemistry, Latin and Greek had a special charm for me, and I studied them thoroughly as I went along so that I was well grounded in these studies. At the suggestion of friends I gave some consideration to preparing for the ministry, and so paid particular attention to Latin and Greek. However, my association with men who were preparing to become preachers convinced me that I could not follow the ministry and accomplish my objective in life. I must say that the mental training and discipline of the mind which the study of Greek and Latin gave have been worth more to me than any other of my studies except mathematics, and I have appreciated this fact more and more with the passing years.

These languages helped me in English, and now I often get the meaning of a word by thinking of its Latin root.

Among the students at Hiwassee were some young men from Louisiana and Mississippi, who were preparing to become lawyers. One of them was the late Robert Broussard, who served Louisiana in the United States Senate with distinction for many years up to the time of his death. I joined these students and studied Blackstone for a while more to broaden my field of knowledge than to make law my profession. I have been invited often to give talks before social clubs and business organizations. Frequently, I have told the following story: "I first studied to be a preacher, but decided that I was too prone to tell big stories to be a preacher. Then I studied Blackstone for a while, and soon learned that I was not adept enough at prevarication to make a successful lawyer. I then made up my mind that I would seek some field where I could tell big stories and tell the truth; later the weather furnished that field." After I had been in college nearly two years my desire to follow some scientific line of work led me to specialize in subjects leading in that direction, but I had no particular professional objective at that time.

During my five years at Hiwassee I went home Fridays after school and assisted with the farm work on Saturdays. Sundays I went to Sunday School and Church, and in the afternoon went back to school. During vacation I performed the work of a farm hand. Harvesting was one of the big farm jobs during vacation. The wheat crop was harvested with scythe and cradle. Stumps which had been left in the field to rot away made it impossible to use mechanical reapers which were just coming into use. Ten men were usually hired by the day to help harvest the wheat crop, five to cut wheat with the scythe and cradle and five to follow and bind the wheat into bundles. Manual labor performed while in college kept me strong and tough. I practiced with the scythe and cradle until I was an expert. The first swath was mine and the other reapers followed me. Each reaper was expected to keep within a certain distance of the one just ahead of him and all would swing their cradles in unison. It was an interesting sight to see the golden grain falling with rhythm into the cradles being swung by six men, then grabbed up quickly with the right hand dropped to the ground and six other men binding it into sheaves as rapidly as the reapers deposited it. Some of the men grumbled, saying that I as leader made them work too hard. Father would reply: "What, you men cannot keep up with a boy!" They could not slow up and drop out, but had to keep on to the end of the row. Some would frequently drop out at the end of the row without waiting for quitting time. There were no organizations in those days which penalized a man for doing his best.

Dr. J. H. Bruner, President of Hiwassee College and his son, Joseph Key Bruner, showed much interest in me and that gave me encouragement. Prof. Joe Bruner often went to Atlanta, Georgia, for short visits to see the girl he afterwards married. I would sub for him in mathematics while he was absent on these trips. Two years before I was to get my Bachelor of Arts degree I had an opportunity to choose a scientific career and be assured of a job as soon as I graduated.

The U. S. Weather Service had been organized in 1871 as part of the Army Signal Corps, and was then about ten years old. General H. A. Hazen, Chief Signal Officer, visualized a weather service with a personnel of college men, who in the event of war would be trained and ready to be commissioned as officers in the U. S. Army. He wrote the presidents of various colleges, and told them of his plan, asking them to recommend yearly from their graduating classes a man who desired to take up the weather service as a profession. A college degree plus good physical condition was all that was required for entry into the service. George A. Martin of Louisiana graduated with the degree of Bachelor of Science, one year ahead of me. President Bruner selected Martin from the first class of about thirty college graduates to enlist in the Signal Corps and go to Fort Myer, Virginia, for instruction in the duties of weather observer. I was to graduate in May, 1882. President Bruner asked me if I wanted the assignment for 1882. I accepted with pleasure for it was just the kind of work I wanted. I graduated soon after I was twenty years old, and father and mother had to give their consent for me to go into the Signal Corps to study to become a weather observer. I then specialized in meteorology and civil engineering until I graduated. I was preparing for the work ahead of me so that I could cross the bridges when I came to them.

In February before I was to graduate, an incident occurred that came near preventing me from getting my diploma, which, of course, was essential to my acceptance by the Signal Corps. Some forty or fifty students of limited means lived in small cottages on the campus and prepared their own meals as I did. An employee of a member of the Board of Directors of the college frequently rode by the campus on a fine steed, taunting the cottage students with "poor white trash" and other insulting epithets.

His route by the campus led through a fenced-in lane about half a mile in length. Mounted on a fine horse he thought he could outdistance us in the event we tried to catch him. We apparently ignored his jibes, but made plans to catch and punish him. A ducking in the creek which passed the college grounds would probably make him realize that he could not get by with such conduct. On a day on which we knew he would pass the campus some of the crowd were sent to the farther end of the lane with instructions to remain hidden until we started after him when they were to appear suddenly, frighten his horse so that it would bolt and start back towards us where we could then halt him. Everything worked out as planned. When the horse found that he was hemmed in the lane he stopped. The rider dismounted and took to the fields and tall timbers. We went after him yelling, "Stop him, catch him and duck him in the creek." We ran through fields and woods for nearly three hours before we were able to catch him. It was in February, the weather was cold, but we were wet with perspiration at the end of the chase. Our quarry was badly frightened because he knew he had it coming to him. The creek was at flood stage. We led him out on a log and pushed him off in the surging stream. He was so scared that he made no effort to swim and went down under the water. When he came to the surface I went into the creek and pulled him ashore. He did not wait to say "Thank you", but ran as fast as his feet could carry him. We thought the affair was over but there was an echo. He contracted pneumonia and came near dying, but, fortunately for me, recovered. Our tormentor never rode by the campus again.

Graduating time was approaching. Rumors were heard to the effect that my diploma would be held up by the Board because I was the leader in the escapade that came near costing a man his life. Prof. Bruner had just returned from a visit to Atlanta, and had not heard of the incident and my connection therewith. I laid the matter before him and gave him all the details. He advised me to keep my own counsel, not to talk with any one about the matter and to tell any one who approached me on the subject that I had "nothing to say". Later he told me that he had looked thoroughly into the matter and that he did not blame us boys for what we did. He said some members of the Board showed considerable feeling against us, but he was certain that a majority would vote to graduate me. When the Board met to grant the degrees he was present and succeeded in getting the full board to approve the granting to me the degree of Bachelor of Arts. President Bruner in presenting me with my diploma congratulated me for high attainments in my studies.

Instructions came from General Hazen telling me to report at the Office of the Chief Signal Officer, Washington, D. C., on July 7, 1882. I was elated, but did not like to think of saying goodbye. Father drove me to Sweetwater, Tennessee, where I was to take the train for Washington. With a sobbing voice he said: "Son, you have been a good boy; I will not worry about you for I will always see you kneeling at night beside your bed to say your prayers." The train came in, I went into the day coach, the conductor gave the "go" signal, the engine emitted a shrill whistle, and I was on my way to a new life and a career.

CHAPTER VI.

STUDENT AT FORT MYER, VIRGINIA

WASHINGTON, the Nation's Capitol, was to me the most important place in the world. I arrived early on the morning of July 6, 1882, and got off the train at the depot where President James A. Garfield had been assassinated the year before. The first thing I saw was the spot and marker where he had fallen and from which he was carried away to die a few days later.

Hotel accommodations were secured near the Office of the Chief Signal Officer. I rested during the 6th, and as this was the first time I had ever been in a large city I was afraid to wander out of sight of the hotel. Promptly on the morning of July 7th, I reported to the Chief Signal Officer for physical examination which I passed 100%. Three other young men reported at the same time, and we were accepted for instruction in the duties of weather observer. The four of us were taken in a two-horse spring wagon up through Georgetown, across the Potomac over the Georgetown bridge, and up through the ridges to Fort Myer which is situated on Arlington Heights, overlooking the City of Washington. Arlington, the former home of General Robert E. Lee, adjoins Fort Myer. The Lee estate is now the National Cemetery where many American heroes sleep the sleep that knows no waking. Here are the monuments of generals and lesser dignitaries and many small marble markers. The small markers indicate the graves of privates, many of whom unobserved and unnoticed had performed deeds of valor, heroism, and sacrifice, surpassing that of some of the generals who have monuments to their memory. Recreation periods often found me resting in front of the Lee Mansion, dreaming of the future and building air castles.

Fort Myer was named for Brigadier General Albert J. Myer. He was graduated in medicine in 1851, was appointed

Assistant Surgeon in the U. S. Army in 1854, and assigned to duty in Texas. His spare time was devoted to devising a system of military signalling by use of flags by day and torches at night. He was appointed to the command of the Signal Corps, U. S. Army, 1858-1860, and was named Chief Signal Officer in 1860. He organized the United States weather service as part of the Signal Corps. The first systematic simultaneous weather observations were collected under his direction by telegraph from 24 stations at 7:45 a. m. November 1, 1870. New Orleans was one of these stations. General Myer did greater things along other lines than those for which he was educated. His career is an outstanding example of what can be achieved by the proper utilization of time.

First Sergeant Mahaney, a veteran of the War Between the States and a fine man, took charge of us on our arrival at Fort Myer. We were fitted out with uniforms and assigned to our rooms in the barracks. Each room had four single beds and thus accommodated four men. There were thirty in our class but five or six of the preceding class were retained to help get us started. The men were divided into squads of ten each which included the squad leader. Sergeant Mahaney put me in charge of the squad to which I belonged. He impressed on me that it was my duty to have the squad fall in at the proper times, march them to and from their class rooms, the mess hall and such other places as he might direct. I was one of the youngest in the class and had no knowledge whatever of military drill. I did not know how to give a single command except "Squad forward March" or "Squad halt". I soon learned that I had to march my squad into the lines formed for inspection, drill, and many other movements. Fortunately one of the boys in my squad had been a cadet officer at the University of Tennessee. I told him of my predicament and he offered to help me through. I placed him at the head of the squad and he would tell me the commands to give in order to get the correct movements and gain our objectives. I learned rapidly and soon felt that I could command an army. I felt very important commanding that squad during my stay at Fort Myer.

Military training in infantry and cavalry tactics formed part of our instruction. The Signal Corps was a cavalry organization and we had lessons in horsemanship. A company of regulars in the Signal Corps, known as permanent men, looked after the stables and horses. However, when we went on cavalry drill we had to groom, bridle, saddle, and care for our mounts and return them clean and nice to their stalls. Some of the men from large cities had never ridden horseback; these men would become badly frightened when we raced around the drill grounds. Some of them would lean forward and put their arms around the necks of the horses, and Mahaney would sing out, "What are you hugging that horse for — straighten up there!" One horse had been a champion trotter in his younger days and he enjoyed the drill. I was a good horseman and always went after him for my mount. Mahaney would place me in the lead and the others would have to keep their distances while circling the drill grounds. I enjoyed making the boys ride with some speed. Mahaney would ride inside the grounds and yell to the men, "Keep your distances— what are you lagging behind for?" Or he would call out, "Cline do not let that horse out so fast." In fact Mahaney enjoyed seeing the boys speeded up.

Military discipline was such as would impress us with our duties. Our equipment consisted of carbines and cavalry sabres, which we were required to keep in immaculate condition. Inspection was held regularly and if our buttons were not polished and our shoes shined including the heels, or if a speck of rust or dirt was found on our equpiment, our week-end leave was cancelled. Our week-ends were usually spent in Washington where we visited the interesting public buildings and museums. Some of the boys went to Fairfax, Virginia, and courted the good-looking girls.

Regular guard mounts were a feature of every day. When my turn for guard duty would come around, I was generally corporal of the guard and now and then sergeant, which meant that I would not have to walk a beat. A staff of Army cooks prepared the meals but we had to serve our turns as waiters on the tables. Two men were assigned daily to this detail. A roster was posted every afternoon showing the details for the following day; if a man failed to be on his assignment he was sent to the guard house, and locked in a cell to make him more careful in the future. On one occasion, Mr. H———, a young man from Alabama failed to note that he had been designated as waiter. I was sergeant of the guard at that time. Mahaney called out, "Sergeant of the guard look and see who is detailed for waiter today." I had to report Mr. H——— as the delinquent. Mahaney ordered me to take a member of the guard with his carbine and march Mr. H——— to the guard house and lock him up. Mr. H——— felt disgraced and wept like a child. After about two hours of incarceration I was ordered to release him. As soon as he could get to Washington he went to see the United States Senators from Alabama and they got him his discharge from the Signal Corps.

Instruction was given in military signaling with flags, torches and the heliograph, and also in the mechanism and operation of the magnetic telegraph and the telephone. We overhauled telegraph apparatus to learn what caused the "click", and strung wires over which that click could be heard thousands of miles distant. The telegraph was developing rapidly; Samuel F. B. Morse, who had given up a distinguished career as a painter of portraits, had prevailed on Congress in 1843 to grant him $30,000 with which to continue his experiments and to construct a telegraph line between Baltimore and Washington, a distance of about 40 miles. The first transmission of messages between these two places created a sensation. Morse at the time of the successful test between Washington and Baltimore offered his patents to the United States Government for the insignificant sum of $100,000. Thirty-eight years later there was a network of wires over the country with men at the keys sending messages to each other thousands of miles away. The telegraph system had already grown to be worth millions of dollars with a future that the most vivid imagination of the day could not visualize.

Alexander Graham Bell was developing the telephone. He held his first public exhibition at Philadelphia in 1876, as a feature of the Centennial Exhibition, just six years before we commenced playing with this apparatus at Fort Myer in 1882. The great future of the telephone was not foreseen by anyone at that time. The Western Union Telegraph Company, with all its rapid growth, refused to pay Morse a moderate sum for his patents. The United States Signal Corps was experimenting with and operating telephones over short distances. The growth and development of the telephone during the last 60 years shows what the ingenuity of man, in a country of free enterprise, can do.

Edison's first crude workable electric lamp was exhibited in 1879. Little did he dream of the great blessings to mankind that would grow from that small crude lamp. It immortalized his name.

Adolphus W. Greely, U. S. Army, took out an expedition in 1881, which had for its object the charting of the Arctic Regions, and the establishment of one of thirteen circumpolar stations which had been provided for. Greely's party became lost. Two expeditions attempted to take relief to Greely and his men and failed. It was recognized that starvation was certain unless they could be reached. Commander Winfield Scott Schley, U. S. Navy, volunteered to lead another expedition in search of the Greely party. This expedition was successful in finding Greely and a few of his men still alive—the majority had died from hunger and exposure. I had expressed a desire to go with one of the early expeditions, but being under 21 years of age, my parents would not give their consent. Youth is eager for adventure and never stops at the suggestion of probable danger. Schley won fame but the men who went with him have been forgotten. I am now, after sixty-three years, glad that I did not go on one of the relief expeditions. I have conquered greater dangers and have successfully crossed bridges fraught with more difficulties, the crossings of which have served to benefit greater numbers of people.

Subjects bearing on meteorology, the taking and recording of meteorological observations and the uses to which they could be applied called for study every minute of our time. Good progress in studies meant early assignment as assistant observer on some station, and this was our immediate objective.

STUDENT AT FORT MYER, VIRGINIA

The instruction was crammed into us so rapidly that many could not keep up and make the required grades. Such distinguished physicists and mathematicians as William Ferrel, T. C. Mendenhal, Cleveland Abbe among others were our instructors.

Stations for observing the weather were being opened in different parts of the country. Assistants who had made good records on stations were selected to take charge of the new stations. To meet the demand for assistants at stations a rigid examination was held; the sixteen passing with the highest grades were to be assigned to stations and the others were to remain for further instruction. I passed sixteenth, and for some reason was one of the first to receive notice to be prepared for orders to go out on station. In our examinations we had been asked to express preference for some scientific line of research related to meteorology which we could carry forward in addition to our regular work of taking and working up meteorological observations. I expressed a preference for some line of meteorological research that would give results beneficial to mankind. Rocky Mountain locusts had been destroying the farm crops in the Plains Region. These insects would swarm through the country, eat up the crops and destroy every bit of green vegetation in their line of march. The swarms were so numerous that the farmers could not cope with them. I was notified that I would be assigned to the Little Rock, Arkansas, weather station where I would have an opportunity to study the influence of weather conditions on the development and movements of the Rocky Mountain locust.

Rocky Mountain locusts and the weather appeared to me an opportunity for worthwhile achievement. I was 21 years old, the world was before me and my enthusiasm was such that I thought I could do any thing that it was possible for man to accomplish. Orders were received to proceed to Little Rock, Arkansas, and report for duty as assistant. The Government furnished me with railroad transportation and an allowance for meals en route. I had no sleeping car accommodations. An army blanket was swung between two seats which made a hammock and I was as comfortable as if I had been in a Pullman berth. The salary and allowances for the assignment amounted to $60.00 a month. Medical services were to be paid for by the Government when no Army Surgeon was available. In addition there was an allowance for clothing, which amounted to about $120.00 a year, but as I did not draw clothing in kind this was retained to be paid to me when discharged so that I would not be stranded.

The instruction of meteorological observers at Fort Myer, was discontinued a few years later under rather interesting circumstances. The instruction given there was of such a nature that it supplemented collegiate education and reinforced the foundations upon which men could build for the future. Military discipline impressed the necessity for neatness, promptness, and punctuality which not only fitted men for the duties of weather observers but aided them in any other line of work which they might elect to follow. While the discipline was of the strictest sort the students were treated as gentlemen. A few years later after I left Fort Myer an officer who had been promoted after having been a tough sergeant in the regular army was assigned to Fort Myer. He was a gentleman by act of Congress. However, he was not capable of recognizing and respecting gentlemanly qualities in other men as had been done by officers while I was there. He was said to have been rough, harsh, and overbearing, and attempted to make the men who were in training for meteorological work do menial service around the post. The young men practically mutinied, and went to their representatives in Congress and got a Congressional investigation. This resulted in the discontinuance of the school of instruction for meteorological work at Fort Myer. Men were then enlisted in the Signal Corps and assigned direct to stations for instruction. It was not practicable to give such instruction on station as had been given at Fort Myer. The 1882 classes at Fort Myer turned out some men who made their marks in the world, among whom I will mention Alexander McAdie who left the weather service to become Director of the Blue Hill Observatory and Professor of Meteorology in Harvard University, and Austin L. McRae who left the service to become Professor of Physics at the University of Missouri. There were others who reached high positions in the weather service. One man out of ten in the classes of 1882 made "Who's Who in America". No such percentage of the men who entered the service after the school of instruction at Fort Myer was closed, attained that distinction.

CHAPTER VII.

ASSISTANT WEATHER OBSERVER AND MEDICAL STUDENT

MY trip from Washington to Little Rock afforded me considerable enjoyment. Something new, something of interest and beauty unfolded before my eyes all the time. The train was ferried across the great Mississippi River and this gave me a thrill; but little did I think that dealing with this river in the distant future, would become an important part of my work in the service of the public. The weather observation station at Little Rock was in charge of Sergeant William U. Simons, an affable and agreeable gentleman. The office occupied well appointed quarters in the Logan H. Roots Bank Building. A room in the same building, near the office, was available for my use. This afterwards proved most advantageous, as no time was lost going to and from the office, and time was a matter of the greatest importance. Observations were taken frequently in those days; the first at 5 a. m. and the last at 11 p. m. My detail opened and closed the work for the day. This detail fitted directly into plans which I afterwards formulated. The official in charge took the observations during the day. Special weather reports were collected during the crop growing season for the agricultural interests. Railroad Station Agents telegraphed reports of temperature and rainfall at 5 p. m. daily. A telegraph instrument in the weather office was connected with the railroad wires and I took the reports as they came in and prepared bulletins for the commercial interests. I assisted in all kinds of weather service work. I was anxious to prepare myself so that I could take charge of a weather observation station.

Meteorological work brought me into personal contact with telegraph officials and their employees. Weather reports from regular observing stations were collected by telegraph through the Western Union Telegraph Company. The observations were received in cipher, and it was part of my work to decipher the reports and make up bulletins for the public and the newspapers.

Alexander Graham Bell was making rapid progress in the development of the telephone. Telephone exchanges were being opened in the larger cities throughout the country. Little Rock was selected as the location for a telephone exchange and stock in the parent company was sold at $100.00 a share, the money to be used for development purposes. Several of the boys with whom I associated bought one share and some of them more than one share. They tried to persuade me to buy a share; I told them that in my opinion the telephone might be used successfully in talking over distances of two or three miles but that would be the limit. I could not visualize that within my lifetime telephone conversations would be carried on successfully around the world. The boys who bought stock and held it became wealthy in a few years. In twenty years one share that cost $100.00 had paid stock and cash dividends to the amount of $50,000.00. I met some of these former boy associates, on whom the Goddess of Fortune had smiled, several years afterwards and they were prominent officials of the telephone company.

Soon after my arrival in Little Rock the State Legislature passed a law making the pronunciation of Arkansas, "Arkansaw".

After finishing my regular duties for the day I found that I had considerable time available which I could devote to something else of a constructive nature. Search for the Rocky Mountain locusts was futile; they evidently learned that I had been put on their trail and disappeared; they did not return to the Plains Region while I was on the Little Rock Station. The investigation of the influence of weather conditions on the habits of the Rocky Mountain locusts was terminated by their movements before it was commenced.

Since the locusts did not give me an opportunity to study the influences of weather conditions on their movements, I looked for some other line of study which I could tie in with the weather. The Medical Department of the University of Arkansas was located at Little Rock just three blocks from the office of the weather service. It offered a three-year course and was rated as one of the best medical schools in the country at that time. Medicine offered a field which could be tied in easily with my work in the weather service. A knowledge of both medicine and meteorology would enable me to make an intelligent study of the effects of weather changes and of climatic conditions on mankind in sickness and in health. Here was a field in which there had been little research, and the subject could not evade me as the Rocky Mountain locusts had done.

Consultation with the medical faculty revealed that my Bachelor of Arts degree would be credited as the first half year in the medical course, so that only two and a half years regular attendance would be required. I was confident that I could prevail on the Chief Signal Officer to keep me on the Little Rock station until I could complete the course in medicine and get my M.D. degree, when I explained to him my object in studying medicine. I started in the middle of the school year 1882-1883. There were some students with academic degrees and there were others who had served an apprenticeship under an experienced regular practitioner. A certificate of apprenticeship under a physician in good standing attesting to qualifications would admit a student to the second-year course. The certificate had to show that the student had made satisfactory advancement in the diagnosis of disease and possessed those qualities which go to make a good physician.

Diagnosis was rated as one of the most important things in medical education in those days. The Professor of Theory and Practice of Medicine often entertained us with stories about the errors in judgment made by apprentices. One is so interesting that I will repeat it here. An apprentice had been serving with a doctor of long experience and high standing as a physician. The doctor with his understudy visited a patient in the country. The doctor told her that she had been eating cucumbers and that she would die. After they had left the house the apprentice asked how he knew the patient had been eating cucumbers. The doctor said that there were cucumber rinds under the bed. He then impressed the apprentice with the importance of keeping one's eyes open. "Observe everything, and always look around and under the bed to see if there is evidence to indicate what the patient had been eating." A few months later the doctor had a patient in the country who was convalescing. This patient lived a short distance off the main road, and the doctor in passing sent the apprentice by to see and report how the patient was getting along. In a few minutes the apprentice, whipping his horse, came at full speed, calling, "Wait, doctor!" The apprentice rode up and reported as follows: "Doctor, she is going to die. She has eaten a horse, for I saw the saddle and bridle under the bed." The professor added, "There are many doctors who do not exercise much better judgment in diagnosis than this apprentice displayed."

My vocation as assistant observer in the weather service and studying medicine in my recreation time were both exacting work, but I found no difficulty in carrying both along successfully; the one gave me rest from the other and I never became tired. My hours of duty were such that neither interfered with the other. When vacation time came and other students went home, I did not stop but continued studying during vacation, taking special lessons from members of the faculty. In this way I kept well ahead of my class and could have passed examinations in all studies well in advance of graduation time. A great deal of time was given to attending patients in the Charity Hospital, and visiting and prescribing, over the City Physician's signature, for charity patients. The faculty was greatly pleased with my application and progress, and particularly with the object I had in view.

Many of the medical students were smokers. I had never smoked and they initiated me into smoking. Before I had acquired the habit I was completely cured of smoking for all time. I smoked a cigarette now and then, one or two cigars had been burned up with great success, and in my own opinion I was a very good smoker. My friends played a practical joke on me. A cigar was given me which was doped, and when I had smoked about half of it nausea set in. It seemed that I disgorged not only what was in my stomach but all of my insides. I was certain that I had been poisoned and was going to die then and there. I was too sick to pray. But I got over it, of course, and have never had any further desire to smoke and no one could prevail on me to try it again. At that time I thought the boys had done me a dirty trick. Now I can appreciate that they did me a great favor. That prank prevented me from becoming addicted to a habit that causes an enormous waste of time and impairs the efficiency of men and women.

Preparedness is of value in more ways than one can imagine. It was fortunate for me that I had studied during vacation time and advanced myself ahead of my class. Thanksgiving Day, November, 1884, I received telegraphic orders from the Chief Signal Officer to proceed by the next train to Fort Smith, Arkansas; the observer there had died of heart failure that morning. The faculty of the medical school was consulted immediately. They assured me that there was no occasion for worry. The Congressional delegation from Arkansas would be telegraphed with request that they call on General Hazen the next day and ask for my early return to Little Rock so that I could take the examinations and graduate in medicine. Immediately upon my arrival in Fort Smith I wrote to the Chief Signal Officer, explained that I was in my last year in medical college and asked that he please return me to Little Rock so that I could finish my medical studies and get my medical degree. I informed him that I was preparing to make investigations relative to the effects of weather and weather changes on man in health and sickness. In a few days a letter came stating that I would be returned to Little Rock, as soon as a man could be sent to Fort Smith to relieve me.

Fort Smith was a town of some importance, but the weather observation station was in a broken down building on the Government reservation, some distance from the business center. Suitable offices were located in the business part of the town. Drawings of the offices selected, and of the roof, showing the proposed location of the instruments were made

ASSISTANT WEATHER OBSERVER

and sent to the Chief Signal Officer, with recommendations that the weather observation office be moved from the Government Reservation into the town proper. I did not remain to move the office but my successor was instructed to put my recommendations into effect.

When I boarded the train to go back to Little Rock the conductor did not take up my Government order for railroad transportation which proved very fortunate for me. There were no dining cars in those days; but there were eating houses where the trains stopped for passengers and the train crew to take meals. When we stopped for supper I was hungry, as I usually am at that hour and before I realized what was happening the train pulled out and left me eating. I soon found out that a freight train would leave in about one hour and reach Little Rock soon after midnight, six hours later than the passenger train on which I had started the journey. I boarded the caboose and showed my order for Government transportation to the conductor; he said—"Freight trains do not carry passengers." When I explained to him that the passenger train had pulled out and left me without giving me notice and that I would have to reach Little Rock by morning, he took the transportation order and noted on it what I had told him. He said: "Take a seat—we will see that you get to Little Rock." The passenger conductor's failure to take up the transportation order saved me from a predicament; I had not received pay since leaving Little Rock and I had only about two dollars left in my pockets. I reached Little Rock in time to report for duty the next morning. That occurrence taught me a lesson, for afterwards I always kept an eye on the conductor at eating stations and when he left the counter I would make a break for the train. I have traveled a great deal and have not missed a train since.

Social affairs took some of my time. I was fond of music and was a tenor in the chorus of a few public performances. While attending school I never permitted social matters to interfere with my work and studies. Women were admired and appreciated for their worth but I never allowed myself to give special attention to any one girl. My objective in life took precedence over everything else; I would not allow myself the opportunity of having a love affair until I had finished my education. However, I learned after a while that my education would never be finished, and I found that some social recreation would aid rather than retard me in achieving my objectives. About eight months before I was to graduate in medicine a beautiful and charming girl just home from school attracted my attention. We were soon very close friends and I admired her more than any girl I had ever known. She would often come with her buggy, after I had finished my work in the office, and drive me to the Charity Hospital or to see some charity patient. Then we would go for a ride and enjoy an interesting talk. We often turned up at her home where I would stay for dinner and spend a pleasant hour. On one occasion a friend of mine came to me and asked, "Are you and Miss ——— sweethearts?" I told him that we were friends and that was all. He remarked that he wanted to marry her but for some reason he could not interest her. His sister was this girl's chum. I told him that the only way to succeed was to keep trying. I was very fond of the girl, but we never broached the subject of marriage, for at that time she was not looking for a husband and I was not looking for a wife. There was an understanding between us so that there was no deception practiced. When I left Little Rock, we exchanged letters for nearly a year during which time they gradually became less regular, then they stopped and the announcement of her marriage to my friend followed. He was a fine man and he had won a prize. She had given me a photograph showing her in all her beauty and charm, which she did not ask me to return to her. This was a friendship that has never faded from my mind. The woman whom I afterwards married was told of this friendship. She took the girl's picture and placed it on the piano where it remained until our house was burned several years later.

On March 29, 1885, I wound up my medical course and the Governor of Arkansas gave me the diploma making me an M.D. April 3, 1885, a letter was received from the Chief Signal Officer in which he stated that he had been much pleased with the manner in which I had handled the affairs of the office during my short stay in Fort Smith. As a tryout for future advancement I would be given charge of a station in west Texas. Orders were received in a few days directing me to proceed to Fort Concho, Texas, to take charge of the station there and complete the transfer of remnants of the military telegraph lines to the Southwestern Bell Telephone Company. The assignment increased my pay to $75.00 a month. Transportation was furnished me by railroad from Little Rock, Arkansas, to Abilene, Texas, and thence by Rocky Mountain Stage Coach, 100 miles, to Fort Concho, Texas. I looked over the latest Rand McNally Railroad Map and there was no Abilene, Texas, to be found. Consultation with the railroad ticket agent revealed that Abilene was a new town which had grown up like a mushroom over night. It was the center of a large and rich cattle industry. One hotel and several saloons had all the patronage they could care for. The ticket agent said:

> "The railroad has just reached Sweetwater, Texas, where Judge Bean has opened his saloon and office as Justice of the Peace. Since you are going there a tenderfoot, you will be interested in an incident which has just occurred. Chinamen are employed in building the railroad; a few days ago a crowd of cow boys went into Sweetwater, got drunk and killed five or six of these Chinese laborers. The cow boys were hailed into Justice Bean's court; he ordered them to report for trial the following Saturday. When the cow boys appeared for trial Justice Bean handed down this decision: 'Gentlemen, I have examined the laws of the United States carefully and I do not find any law which says that a white man shall be punished for killing a Chinaman; you are free.'"

Every person I met had a story to tell of the treatment the cow boys of that region dished out to a tenderfoot who had the misfortune to find himself among them. I was told that well-dressed men often had their hats shot off their heads and their good clothes pulled from their backs. There was some truth in this, for such things were done. During my last year in medicine I did work at the Charity Hospital, and with some practice I felt like I was a full-fledged doctor. I dressed in the style set by the foremost physicians of Little Rock. I donned a Prince Albert brown beaver suit, silk hat, kid gloves, and carried a cane. I was on the top of the world and ready for anything except to meet a Texas cow boy. However, I would have to meet the cow boy, and I realized that my style of dress must be changed. I had with me a suit of clothes I had worn when on the farm in Tennessee, which I had kept to wear on fishing trips. (Yes, I went fishing with my girl friend and her cousins occasionally and we caught fine trout, which we broiled and ate on the banks of the stream.) The clothes which I had been wearing while prescribing medicines for charity patients were packed in my trunk so that should the cow boys examine it they would not be found, as there was a false bottom to my trunk. Disguised as a farmer in old clothes and with a farmer's manners I started to Texas to meet the cow boys.

Trains did not run on regular schedules in those days, especially over newly built roadways. Heavy rains had fallen over western Texas and many of the bridges over the small streams had washed out. We were delayed frequently until repair trains could come and rebuild the bridges, or replace a washed-out stretch of track. Abilene came in sight late in the afternoon, and the first thing I noticed was a large congregation of cow boys with their high boots, large spurs, big hats, and with pistols in holsters hanging from their belts. The stage coach was not due to leave until the following morning, and the thought of remaining in Abilene all night with such a fierce looking crowd of cow boys was anything but pleasant. I could not get a room in the hotel, but the railroad agent, to whom I carried a letter of introduction, got me a room for the night over a nearby saloon. When I reached the saloon a porter was washing up something red on the sidewalk, and I remarked, "That looks like blood". The porter replied, "Yes sah, four cattle men fit it out here dis 'eboning, and all four ob em is dead." These were not cow boys but rich men, large cattle owners. I never learned the cause of the duel, but cattle rustling was probably at the bottom of the trouble. I afterwards owned the home of one of these men. My head did not rest easy that night; the tramp of cow boys and the shooting of pistols made it a night of suspense.

Morning finally came, bright with cheerful sunshine which portended a pleasant journey over the plains. The stage coach, four in hand, pulled up at the depot. Four passengers were waiting, all bound for a through trip. A young matron was going to join her husband who was opening a drug store in the new town of San Angelo which was on the near side of the Concho River. A boy about 17 years old, whose parents were sending him to the ranch of an uncle near Ben Ficklin about fifty miles up the river from Fort Concho. A big fat fakir, a patent medicine man, going to the new town of San Angelo to peddle his wares. Then the weather observer en route to Fort Concho. Some Mexican burros were grazing a little distance from the road and the boy sang out, "Look at the jack rabbits!" The schedule of the stage was 100 miles in one day and the horses were changed every 30 miles. We were scheduled to reach San Angelo late that afternoon, but a stream which under ordinary conditions was forded by the stage coach was swollen by a flood when we reached it, and we could not cross. The driver informed us that we would have to spend the night there, and wait for the stage coach which would come in from San Angelo the next morning. Then we, with our luggage, would be ferried across the stream in a skiff kept for such emergencies. We had eaten supper at the stage station about ten miles back, the nearest habitation, but there were no accommodations for passengers at the river side. We let the woman sleep in the coach, and the men found a tuft of grass or a rock for a pillow during the night on the wild prairies. About midnight I heard the music of a rattle snake which frightened me badly, in fact so much that I ran and jumped on top of the stage coach and scared the woman into hysterics. She thought the Indians, who appeared in that neighborhood sometimes, had attacked us: I remained on the coach until morning. Soon after daybreak the stage coach from San Angelo appeared on the opposite side of the stream. We were ferried across in the skiff and were soon on our way. I stopped at the hotel in San Angelo, about one mile from Fort

Concho, which was on the other side of the Concho River, at times a very important stream.

Rainy weather had caused the prairies to produce a carpet of flowers such as words will not describe. The flowers rolled in the wind like vari-colored waves. In fact Texas greeted me with flowers—flowers to the east, flowers to the south, flowers to the west, flowers to the north, flowers everywhere—the most beautiful vision in nature my eyes have ever beheld.

CHAPTER VIII.

IN CHARGE OF THE WEATHER OBSERVATION STATION AT FORT CONCHO.

FORT CONCHO, Texas, was on the fringe of a region marked on the maps of that time as the "Great American Desert." Geographers evidently had not seen that part of the country after a season of general rains during the winter and early spring, or they would have labeled it the land of promise. The explorers had probably been through there during a dry period. A carpet of flowers, waving in the breezes, bespoke the fertility of the soil which now appeared only to be waiting for the plow, the sowing of the seed, and the reaping of bountiful crops.

The Concho River was quite a stream in the wet season but in the dry months long stretches of its bed were free of water. But there was at all times a large volume of water flowing through the shale under the dry river bed. Some stretches of the river bed, as much as a mile in length, where the shale was mixed with sand, had washed out to depths of ten to twenty feet and these were always full of water and stocked with an abundance of fine fish.

The headquarters for that section of the United States Military Telegraph, had been located at Fort Concho. Telegraph lines had connected the military posts of that region and formed part of the strategy for combatting the Indians. The Indians having moved to the Indian Territory and New Mexico, the Southwestern Bell Telephone Company had moved in and bought up the United States military telegraph lines which they were operating commercially, as well as handling Government business. In addition to my duties as weather observer I had to complete the transfer of the telegraph equipment to the telephone company or to United States military posts on the Mexican border. A cottage located near the Fort Concho reservation was occupied as the weather observation station and sleeping quarters. I took my meals at the hotel in the new town of San Angelo. Weather observations which were telegraphed three times daily to Washington were filed with the telegraph office in San Angelo. Lee Stockard was the telegraph operator. He resembled some one I had known so strikingly that I often was tempted to ask him something of his previous history. However, I had been warned not to inquire into the previous history of men I met in the far west, but to take them as I found them. After giving him my weather message I would remain occasionally and join Stockard in a game of chess. He was a likeable fellow, a few years my senior.

Many stories were current about happenings during the early settlement of that section. I can not vouch for this story but I am inclined to believe that much of it is true. Up to a short time before my arrival, new-comers from back east were given an invitation to attend a meeting with those who had assumed leadership in the new community. They were informed that the meeting was to enable them to get acquainted for the purpose of mutual help. The leader or chairman of the meeting would give a talk on obedience to the common law, as there were no organized courts. The new arrivals were told that the men in the community were honorable and substantial, but some of them had come to the far west to escape complications back east. Each immigrant was questioned concerning what he might have done back east that had caused him to come west. He was assured that this proceeding was of a confidential nature and was for the purpose of helping him in the event of trouble in the future. This method enabled the citizenry to weed out undesirables. The community grew rapidly, and many of the new arrivals were able to say that they had come west to grow up with a new and promising country and that there were no complications that had caused them to change residence. New arrivals were accepted subject to their future conduct. These people were friendly, generous to a fault and they would go a long ways to help a law abiding person. But woe to the person who failed to obey the law.

Mid-August, one of the hottest months of the year, was interesting to me as I recorded the highest temperatures I had ever experienced. There had not been a drop of rain since my arrival in April. In fact no clouds had appeared anywhere on the vast expanse of blue sky. The beautiful flowers had faded, and the prairies were brown and bleak. However, there was still good fishing. The water holes in the Concho River always received a good supply of water from some seemingly unknown source.

One day as dinner time was approaching and I was walking leisurely across the foot bridge spanning the high banks of the Concho, I suddenly heard a roaring noise upstream. To my amazement there was a head of water some fifteen or twenty feet in height bulging down over the dry bed of the river. Some fifty yards above the foot bridge a man with two women was driving across the dry bed of the river where a cattle trail had worn down the banks. The head of water was traveling so rapidly that it caught them before they could get out of its way, and they were carried down stream and drowned. The water rose rapidly and was soon nearly up to the footbridge and I hurried across to the San Angelo side of the river. When I reached the San Angelo bank I looked upstream and saw Mexicans and others taking large fish out of the water where it had spread out over the cattle trail. I was always interested in fishing and walked up to the cattle-crossing to find out how they were accomplishing the feat. A fish some two feet in length, unable to use its fins, drifted out to near where I was standing. I reached down to pick up the fish and lo and behold, my hand plunged into ice water! This icy water had chilled the fish until they could not swim and could be gathered in with the hands. No one there had seen anything of the kind, and all marveled that there should be a river of ice water in Texas in mid-summer. What caused the icy flood to come pouring down the dry river bed in summer? This was a question asked by everyone. There were several scholarly men in the community who theorized on the origin of the ice water, but none was correct. About ten days after the occurrence of the flood of ice water some people came down from Ben Ficklin, a place about 50 miles up the Concho River from San Angelo, and we learned all about the cause of the flood. They told us that a hail storm of great severity had visited that section and covered an extensive area. The hail stones, some of which were as large as ostrich eggs had killed hundreds of grown up cattle. So much hail fell that erosion gulches were filled and the hail was three feet deep on the level ground. This hail melting in 160 degree August sun caused a river of ice water in Texas that froze the fish till they could not swim so that they could be gathered in with the hands.

Here I thought was my opportunity to make a record with the officials in Washington. A paper on the subject, describing every detail, was prepared and submitted for publication in the United States Monthly Weather Review. The editor of the Review, to my surprise and chagrin, wrote me that the story could not be published, and remarked that it is evident that my association with the disciples of Blackstone had developed an unusual capacity for exaggeration. The contribution was never published.

Great hailstorms have been reported from different parts of the world, but none of them caused a river of ice water in summer that chilled the fish so that they could be caught with the hands. In Moradabad, India, on April 30, 1888, hail stones of enormous size killed more than 230 persons and live stock in great numbers. At Yellow Stone Valley, in the United States on June 30, 1877, hailstones as large as oranges killed a great number of ponies. At Whitehall, Illinois, June 2, 1881, hailstones nearly as large as goose eggs fell to a depth of twelve inches. In Audubon, Cass, and Henry Counties, Iowa, on June 12, 1881, hailstones as large as a man's fist fell, and in places were two to three feet in depth. Hailstones which measured 17 inches in circumference, some of which weighed one pound and twelve ounces, fell at Dubuque, Iowa, on June 16, 1882. Hailstones killed a number of horses, cattle and sheep in Lincoln and Miguel Counties, New Mexico, on June 7, 1886.

The Honorable J. Stodard Johnson of Louisville, Kentucky, and General John Sayles and his son Henry were interested financially in Abilene, the new and growing town on the Texas and Pacific railroad. They also owned a large part of the land in that section of the State. A weather observation station would be a good addition to the new town. The weather records would be of value in determining to what extent the land could be used for farming operations. They took the matter up with the Chief Signal Officer, and he told them that he would move the weather observation station from Fort Concho to Abilene. Notice was sent me to prepare to close the observation station at Fort Concho and get ready to go to Abilene to open a First Order Weather Observation Station. This meant that all self-registering instruments used by the weather service would be used on the Abilene station.

Charles F. von Herman, an assistant weather observer in the Signal Corps, had been assigned for signal duty temporarily with the regular Army. Geronimo and other Indians in New Mexico had been on the war path, and the wily Indians under the leadership of Geronimo were hard to capture. The heliograph, telegraphing with flashes of sunlight, was von Herman's detail, and he aided materially in the capture of the Indians. Geronimo said he could outgeneral the Army until they commenced talking with the sunlight of the Great Spirit after which he could not escape capture.

I later met Geronimo at a fair in Omaha. He was a tall handsome Indian, and evidently a very intelligent man. After Geronimo was captured, von Herman was sent to Fort Concho to pack up and ship the instrumental and other equipment to me at Abilene to which place I proceeded by stage coach. The journey of one hundred miles was made without delay.

CHAPTER IX.

IN CHARGE OF THE WEATHER OBSERVATION STATION AT ABILENE.

A mushroom growth in population brought all kinds of businesses to Abilene. Business houses built of brick, two and three stories in height, already covered several blocks. Offices suitable for the new weather observation station were found on the second floor of a three-story building, and were leased for a term of years. The law offices of Sayles and Sayles were located near by, and this proved of material help to me in many ways. The elder Sayles had been a General in the Confederate Army and was the author of "Sayles' Digest of the Laws of Texas", an authority everywhere except in Justice Bean's court which had been moved to the west of the Pecos River.

The instruments and other office equipment arrived soon after I had secured proper offices. Observations of the weather instruments were made several times a day, the last being at 11 p. m. Observations were sent by telegraph to Washington and other large cities twice daily. A good working library was furnished the station. Among the books were reports of the Secretary of War and other publications which contained the complete daily observations of temperature and rainfall recorded by U. S. Army Surgeons at the several military posts from the date of their establishment until they were closed. My first impression that the fringe of the "Great American Desert" was a land of promise was strengthened as the years passed. My recreation time was devoted mainly to the study of the rainfall and temperature records of that region. The first study that I completed was the "Precipitation and Its Sources in the Southern Slopes", which was published in the official bulletin, the United States Monthly Weather Review. The southern slope of the Rocky Mountains embraced Northwest Texas. The publication of this contribution coincided with the closing of my fourth year in the weather service. I had occasion to read this study recently, and now sixty years afterwards I do not find that I could make improvements in the deductions.

Sayles owned the Abilene Daily and Weekly Reporter and Alf. H. H. Tolar, a fine gentleman, was editor and publisher. Articles on the climatic conditions of Abilene and the surrounding country were written for publication in the Reporter during my recreation time. General Sayles would look over my writings and point out to me where style and verbiage could be improved. My association with him helped me in many ways.

Hart ———, who operated a jewelry store, occupied a room just across the hall from my room. Occasionally we spent an hour at night in a game of poker but without an ante. I was adept at the game and soon played a better hand at poker than Hart ———. We were playing a close game on one occasion, he stopped and laid his cards on the table and said:

"Cline, you are only a boy starting life with a bright future before you. Every one around here thinks I make my money in the jewelry business, but I do not. The jewelry business is merely a blind; I am a professional gambler. I make my money playing poker with the rich ranchmen. I am successful now, but some day I will like every other gambler, lose everything I have. When poker playing gets a hold on a person it is difficult to stay out of the game. When you are winning you will not, in fact, can not quit. When your cunning fails you and you are losing, you always think your luck will change and you can win back what you have lost; you continue losing until you have lost your last dime. I advise you to quit playing poker; do not risk playing social games for they are dangerous. If you play poker it will get such a hold on you that it will ruin your career."

His advice was followed and I have not played a game of poker since that night. Some years later I was on an inspection trip and stopped in Oklahoma City; my eye caught the name Hart ——— on the window of a jewelry store, a palatial affair. I went in and asked for Hart ———. They told me that he did not own the store and his whereabouts were not known. Some thirty-five years ago a man, straight as an arrow, with piercing eyes, a determined face, and black hair streaked with grey, came into my office. He extended his hand and I knew that I had seen him somewhere but I could not place him. When he told me his name I remembered him well. I then thanked him for the advice had had given me, for I had seen many men ruined by playing poker. He told me that he had made and lost two or three fortunes and that he was again down and out. Hart ——— stated that he was going to some small town to open a cheap jewelry business and try to recoup his fortunes. I have heard nothing from him since that visit.

General Sayles called me into his office and congratulated me on my twenty-fifth birthday. He informed me that Tolar could not make enough out of the Reporter to support his family and that they were looking for some one to edit and publish the paper. He said:

> Dr. Cline, you do a great deal of writing and we would like to arrange with you to publish the **Daily and Weekly Reporter.** It will not be a money-making venture, but you will get experience worth something.

The proposition was taken under advisement. There was an opening for a job printing shop in Abilene. Job printing for business houses in Abilene was being done in Fort Worth and Dallas. A tramp printer, assisted with a guarantee by one of the banks, ordered and had shipped to Abilene a Gordon press and the equipment to run a job printing business. The job printer disappeared and the printing outfit was still unboxed in the rear of the bank. I found that I could buy the outfit for about half its value. Another printer named Neely was looking for work. He wanted me to put up the money, buy the printing outfit, and manage the business, he to do the printing for half the profits. General Sayles let me put the job outfit in the office of the Reporter. I took over as editor and publisher of the daily and weekly Reporter, as something to do in my recreation time. The paper was democratic, but it was supported by both the cattle and sheep ranchers. One crowd wanted free trade and the other a protective tariff. Friends told me that I would have trouble because my clientele included some dangerous men. I was advised to keep a couple of loaded shot guns near the editorial desk—that a gun was a good bluffer. A 45-caliber, hair trigger, winchester rifle which I used for shooting prairie dogs, jack rabbits, and ducks, was kept standing by my desk. I never had occasion to use it except to shoot game. I had prepared for a duel that never took place. Tolar's son was employed as printer to operate the old Washington press on which the Reporter was printed. The Reporter paid expenses. Another paper, ABILENE WEEKLY NEWS, was published by a Mr. Gilbert, a very fine man.

Job printing proved to be a money making business. A monthly paper, THE WEST TEXAS REVIEW, was published. The **Review** had for its object the publication of the temperature and rainfall records of that region so that people who contemplated coming to that part of the country would be correctly informed regarding the important climatic features which determine the crops that can be successfully grown in a locality. When all the records for the several observation stations had been printed, the **Review,** having served its purpose, was discontinued. Neely offered to buy my interest in the job printing plant at a handsome profit. I sold the job printing business to him and endeavored to make the **Reporter** a better newspaper.

Cumberland Presbyterian was the religion in which I had been brought up, but I found myself a regular attendant at the Baptist church. Rev. George W. Smith, the pastor of the church, had living with him his wife's niece, Cora May Ballew. She was a beautiful, brilliant, and cultured girl. She was a fine musician and the church organist. She held more attractions for me than any woman I had ever known. I wooed her and she promised to be my wife. We were married on March 17, 1887.

About this time I met a man with a familiar face, and I called out to him, "Hello Stockard!" The man straightened up and said, "My name is not Stockard. I know you. I was in college with you, and visited you in your home back in Tennessee. My name is Murphy." He looked so much like Lee Stockard, who was telegraph operator at San Angelo, that I had mistaken him for that person. He told me that while teaching school back in Middle Tennessee, he whipped one of the pupils and as a result had an altercation with the boy's father and killed the man in self-defense. He left Tennessee and had been residing near San Angelo where he owned a ranch. He had recently been acquitted of the murder charge. I invited him to visit me at my home, but I never saw him again. Resemblance is sometimes so striking that it leads to errors in placing people.

My wife aided me materially in editing and publishing the **Reporter.** Our first child, Allie May, was born on December 10, 1887. As soon as my wife was strong enough she resumed work on the editorial staff of the **Reporter.** This gave me an opportunity to devote some of my spare time to other things.

The State Fair, which was to be held at Dallas, offered a banner to the county sending in the best agricultural exhibit. Henry Sayles asked me what I thought of having Taylor County compete for the banner, and if I would undertake the job of preparing the exhibit. He was advised that we stood a good chance of winning the banner. A good growing season presaged bountiful crops. Furthermore the older counties would rest on their usual agricultural reputations, and no special efforts would be made to put on a fine exhibit. Some Chinamen had gardens along the creeks where they could irrigate their crops when there was a shortage of rain. The vegetables they produced were of the finest quality and the greater part of their output was shipped to Fort Worth and Dallas. Specifications did not eliminate crops produced by irrigation. We believed that Taylor County, even though located on the fringe of the "Great American Desert", could send an agricultural exhibit that would carry off the banner. Sayles offered prizes, four for each product in order to get at least that number of persons to compete for the prizes. The prizes were $50.00, $25,00, $10.00, and $5.00, for the four finest water melons and so on for other products. From the four we would select the best for the exhibit. The preparation of the exhibit was given the most careful attention. Perishable products were put in glass containers with preserving liquids and sealed. One glass case contained a water melon that weighed 150 pounds. Taylor County won the banner.

Crime was severely punished in that section of the country. As is frequently the case much of the crime was caused by the prevalence of many saloons. The **Reporter** took up the fight for local option, and put Taylor County under the local option law which closed the saloons. County Sheriff Cunningham was a most fearless and just law enforcement officer. He could pull his gun and shoot as quick as a flash of lighting. He brought in two bandits by himself once. The bandits were bound on one horse. When he took off their cords to put them into the jail they broke away, mounted the horses and tried to make their escape, but Cunningham's unerring aim left the two bandits stretched out dead on the street. With local option there was little need for such an efficient law enforcement officer and the State took him as Superintendent of the State penitentiary.

During my stay at Abilene I kept up my studies of the effects of weather changes and climate on mankind, which at that time was the real objective of my life's work. A good working library of old books relating to the subject was brought together. Statistical matter bearing on the subject was collected for use in future studies.

Marvelous stories were told about sudden temperature changes in that section of the country. One of the most far fetched—and yet possible—stories concerned an incident that happened a few years before the railroads made their advent into that region. Freight for the Far West was hauled by ox teams. A freighter was hurrying his team forward one hot morning, when at a point a short distance west of where Abilene now stands, a sudden heat wave swept down on him. One of his oxen could not stand the heat and died. The hide was valuable, so he walked back a few miles to the freighters' way-station to get a knife to skin the ox. Before he had time to return to his wagon a norther had swept down over the plains and the other ox had frozen to death.

Hot waves of air have occurred in that region, fifty yards to one-fourth of a mile in width, that burned vegetation to a crisp.* An animal not in good physical condition when exposed to such a heat wave might die. A sudden change from the heat wave to severe cold could have chilled the other ox and caused its death.

One April morning, which was unusually warm, I drove some distance into the country to shoot prairie dogs and jack rabbits. I did not wear a coat and had no lap robe. When it was time for me to return to town I saw a cloud having the appearance of a tremendous roll of black wool bulging down over the prairies. I recognized it as a messenger announcing the approach of a norther, and the rapidity of its movement indicated that it would be decidedly colder in a very short time. I was wet with perspiration and my horse was in a lather of sweat. I hurried back to town as rapidly as my horse could take me. When we reached town I was frozen almost stiff, and there were icicles on my horse's nose and hair. If we had been caught in one of those narrow heat waves immediately preceding the cold our metabolism or heat control processes might have been taxed to an extent beyond their capacity and death might have resulted.

Tornadoes are not of frequent occurrence in northwest Texas but I happened to see one at too close range for comfort. While driving along a roadway formed by wire fencing, about three miles from Abilene, I heard a terrific roaring noise behind me. Looking around I saw a tornado with its funnel-shaped cloud swirling in the air and traveling parallel to the road and only about two hundred yards away from the road. The tornado was traveling at a speed of about ten miles per hour while the wind velocity in the swirling storm was so

---

* See SUMMER HOT WINDS ON THE GREAT PLAINS by Isaac Monroe Cline, Bulletin, Philosophical Society of Washington, Vol. XII, pp. 309-348, Plates 4-6, Washington, D. C., 1894.

great that it could not be measured. I stopped and watched the tornado so that if it should take up a course towards me I could turn and go in an opposite direction and escape. As it crossed Lytle Creek it carried the water up in a solid stream. It tore mesquite trees up by the roots and whirled them about in the air like feathers. The column connecting the funnel cloud with the earth became over extended and parted about one-fourth of a mile above the ground when the water which was carried up out of Lytle Creek poured down on the earth and washed out a large hole in the ground. The funnel cloud continued to swirl in the clouds above the earth and again reached down to the ground about two miles farther to the east.

Tornado is the name which was applied some two centuries or more ago to the violent thunderstorms that frequent the western equatorial coast of Africa. The tornado in this country is a small but violent storm attended by a funnel-shaped cloud which forms, with thunderstorm conditions, in the front part of extra-tropical cyclones which travel eastward across the United States. Tornadoes do not occur with all cyclones, but several tornadoes occur sometimes during the passage of one large cyclone. The diameter of the tornado varies from a few hundred feet to a mile or more in width, and the length of the track of the tornado varies from a few hundred yards to several miles. When the wind in the cloud region, which carries the cyclone forward, has a greater velocity than the wind near the earth the upper part of the tornado is carried forward so rapidly that it breaks away from the lower part and leaves it to die out, which it does in a few minutes. The funnel part of the tornado continues swirling in the cloud region for a short time, except that occasionally after skipping over some distance it extends down to the earth again with as much force as it displayed at first. Tornadoes generally travel forward with a speed of 10 to 30 miles per hour, the velocity of the winds in that part of the cyclone where they occur. No structure within its path has withstood the winds which attend the tornado. Iron bridges have been bent and twisted into every conceivable shape. Buildings have been torn into fragments. Pieces of planks have been driven through tree trunks. Straws have been driven through a plank like a bullet shot from a gun. In one instance some chickens in a basement under a building which was destroyed by a tornado had all the feathers removed from their bodies. This probably resulted from the extremely low pressure in the center of the tornado. The tornado and the water spout are of the same general nature—a tornado over land and a water spout over water. The correct name for all storms attended by a funnel-haped cloud is **tornado,** and not **cyclone,** the term frequently used in referring to such storms. Cyclones are large storms with diameters of 100 to 1000 miles. As stated above the tornado occurs in the front part of cyclones, usually in the right front quadrant, when there are intense unstable conditions. I have not been able to learn who applied the name **tornado** to the storm with the funnel-shaped cloud. It was apparently used because of the violence of the storm.

The time for the election of a representative to the State Legislature was at hand. Alf H. Tolar was Sayles' candidate, and was supported by the **Reporter.** He was elected by a good majority. California and Iowa had organized local State Weather Services cooperating with the U. S. Weather Service. The advantages which Texas would gain by having such a service was placed before General Sayles, and he drew up a bill providing for the organization of the Texas State Weather Service. Tolar introduced the bill in the House of Representatives, made some good speeches and it passed the lower house but was still on the Senate Calendar when the session adjourned. Copies of the bill and the debates concerning it in the legislature were sent to the Chief Signal Officer. The **Reporter** carried editorials from time to time pointing out the advantages which would accrue to the State of Texas from such a service, and these were sent to General Greely. General Hazen had died and Adolphus W. Greely of Arctic fame had been promoted to Brigadier General and made Chief Signal Officer. General Greely was progressive and he was pushing the weather service to the front.

When the Texas Legislature adjourned without passing the bill creating a Texas State Weather Service the fact was reported to the Chief Signal Officer. He wrote me that he would have a Texas Section of the United States Weather Service organized with headquarters in Galveston and that I would be sent in charge of the Galveston station with instructions to organize the Texas Section of the Weather Service. In March, 1889, I turned the Abilene station over to my successor and went to Galveston to carry out the instructions of the Chief Signal Officer.

TORNADO OFTEN INCORRECTLY CALLED CYCLONE.

The spout connecting the funnel with the earth became separated from the funnel a few minutes later and the lower part died out immediately. The funnel extended down from the cloud to the ground about two miles to the east and wreacked destruction again.

CHAPTER X.

IN CHARGE OF STATION, SECTION DIRECTOR AT GALVESTON.

GALVESTON is situated on Galveston Island, which is about 28 miles long and between 3 and 4 miles wide. The elevation of the Island ranges from sea-level to 10 feet above. Sometime about 1815, after the Battle of New Orleans, Lafitte and his fellow pirates opened headquarters on Galveston Island and operated from there for a few years. Galveston was first settled in 1837, just 52 years prior to my arrival. The population in 1889 was about 40,000 persons.

Texas produced a large percentage of the country's cotton crop, and Galveston was one of the most important cotton markets in America. General Greely in a letter to the Secretary of the Galveston Cotton Exchange said that an energetic young man who had displayed extraordinary ability was being sent to Galveston to organize the Texas Section of the United States Weather Service. General Greely understood Texas and its needs, for he had superintended the construction of military telegraph lines in that great State in 1878-1879.

The assignment to Galveston afforded me great opportunities for the utilization of my recreation time, although the duties devolving upon me were varied. I was executive officer directing the work of the Galveston station, the cotton region service for Texas and section director of the Texas Section of the United States Weather Service. Four assistants, a large number for a station in those days, were assigned to help carry on the work. The organization of the Texas Section of the United States Weather Service was soon accomplished. The Galveston Cotton Exchange cooperated by defraying the expense of printing the monthly Texas Weather Bulletin. The bulletin carried the complete weather data recorded at the several stations in Texas, a feature of great climatic and commercial value. I edited this publication, and my first studies of the effects of weather changes in causing sickness and death were printed in the Texas Weather Bulletin. I was a registered medical practitioner in Texas, a member of the Texas Medical Association, as well as the American Medical Association. I was instructor in Medical Climatology in the Medical School (located at Galveston) of the University of Texas. This was 55 years ago, and so far as I know no other medical college had taken up the study of Medical Climatology. My course of instruction was ready for publication in book form when it was washed away in the hurricane of 1900. My work at the medical school was on my recreation time and did not interfere in any manner with my official duties. I carried on both lines of activity successfully. When the public needed my services I never counted hours.

General Greely did not forget his friends when making appointments to the position of assistant observer, but he would not allow political influence to sway him in dealing with officials on station. In his relations with them he was gentle, but firm. During his first year as chief of the weather service he had many observation stations inspected, and delinquencies of one kind or another were found that resulted in the dismissal of something like one hundred employees that year. The following instance illustrates his attitude towards politicians. An official in charge of one of the most important stations of observation in New England was unusually popular, especially with the ladies. The nude in art appealed to him so strongly that it got the best of him. He was summarily dismissed from the service without the reason for his dismissal being made public. The two Senators from that State called on General Greely and demanded his reinstatement. The General told them that this man's conduct had been so disgraceful that he could not be reinstated. The Senators insisted, and asked for an investigation. General Greely told them that the official had been dismissed because he had been found guilty of making photographs of nude girls in the weather office, girls of prominent families in the city, and that an investigation would bring disgrace on the families of the girls. The Senators withdrew their demands for an investigation. The names of the girls were never made public, and they in all probability later became prominent and respectable women. Another observer in the midwest was addicted to poker playing. He lost all his money, and gambler-like, he had to have money. He went to a local pawnbroker and pawned the instrumental equipment of the observation station and the pawnbroker moved the equipment to his place of business. When the inspector went into the weather office there were no instruments there. He found the instruments set up in the pawnbroker's store and the observer was taking observations in the pawn shop, instead of at the weather service office. In the Rocky Mountain district where good fishing was tempting, the observer occasionally took the observations for a whole week in advance; he would write out a message and guess what the weather would be for each day he expected to be absent. He would then give the bunch of telegrams to the telegraph operator with instructions to send one report each day, commencing with the top sheet and follow through daily. The inspector dropped in, and, not finding the observer, who had gone fishing, he went to the telegraph office where he found the fake observations made up for several days in adavance. Another observer was sent to that station immediately.

About this time, a man belonging to a prominent army family who had been forced to resign his commission in the army because of his over-indulgence in strong drink, was sent to the Galveston station for duty as assistant observer. General Greely and a brother-in-law of the man, both wrote me personally and told me of the man's history. He had taken the Keely Cure and had not touched whiskey for two years. They asked me to be positive and firm in my dealings with him, and said they felt sure I would do my best to keep him on the water wagon. The ex-army officer remained sober and rendered good service for something like a year or more. He then started to get drunk occasionally. The Mayor of Galveston was prevailed on many times to let us take him out of the calaboose so that he would not have to go to trial. We would send him home to sober up, and he would promise never to touch whiskey again. However, when pay day arrived he would get drunk again. Soon the dives, which he frequented, learning that they could get their pay at the end of the month, gave him credit, and then he was drunk nearly all the time. I frequently made talks to business organizations and social clubs. On my return from one of these luncheon talks I found that the weather office had been converted into a bird store. There stood the assistant observer drunk as a lord singing out as he reeled, "Birds for sale, hic hic. Fine birds for sale cheap, buy now, hic—hic—hic!" One of the other assistants took him to his room and put him to bed to sober up. The bird cages had the name of the dealer on them and we telephoned him to come and get his birds. He had sold the birds on credit and when he learned the facts he was glad to get them back. Our army friend finally reached the stage where he was totally incapacitated for work. The facts were reported and his dismissal recommended. Jefferson Davis died while his dismissal was pending. The Cotton Exchange, in which the weather observation office was located, had the United States flag at half mast in honor to the President of the Confederacy. The assistant observer came reeling into the office so drunk that he could hardly stand up, yelling at the top of his voice, "Flag at half mast for Jeff Davis, come, let us pull it down!" He was told that this was none of his business. He went to the telegraph office and sent the following message: "Greely, Washington, flag at half mast, and I am in the minority", signing his name. He was dismissed from the service. When I handed him his dismissal papers with a check for the amount due him, about $400.00, I told him to buy a ticket and go to his people in Washington. Two weeks later a telegram from them asked if we knew his whereabouts. We found him in a low dive with no money. We took him out, sobered him up, and told him that unless he went to Washington at once we would have him locked up for vagrancy. He began crying, and said that he could not go because the dive people had all his money. We bought his ticket to Washington, gave him a few dollars to get something to eat en route, and placed him in charge of the train conductor to whom we explained his plight. A graduate of West Point and a brilliant man, whiskey had wrecked his career and life. We never heard from him after he left Galveston. However, there was an echo. The hard-boiled officer, of whom I have previously spoken, was executive officer in the office of the Chief Signal Officer. He wrote me a letter in which he tried to make it appear that I had recommended the man's dismissal because of the flag incident at the time of Jefferson Davis' death. I laid the matter before General Greely personally. The hard-boiled officer apologized, claiming that he had acted under a wrong impression. Afterwards I found that one of the dive bums had written the Central Office, alleging that the flag incident was the cause of the man's dismissal.

General Greely, when a lieutenant in the United States Army during the Civil War served under the young and brilliant Major H. C. Seymour. Greely stayed with the Army. Seymour went back to civil life, but things did not go well with him. He drifted far below his station in life as the result of drinking whiskey. When Greely was made Chief Signal Officer in charge of the weather service Seymour applied for a position as observer. He failed to make good at two stations and was sent to Galveston for a final trial. Greely wrote me, explaining the situation, and told me not to allow his former relations with Seymour to influence my actions in dealing with him. Seymour was highly educated, but a mystery man. Nothing could be learned about his antecedents. He had left England when a young man; joined the United States Army when war broke out between the States and rendered distinguished service. Just another derelict as the result of strong drink. He gave fairly satisfactory service for a while, then apparently made up his mind that the Government owed him a living. He would go to his desk and just sit there all day long, not doing any work. He was told that he had been assigned to the Galveston station to perform the work of an assistant, and that unless he went to work I would be compelled to recommend his dismissal. He replied, "General Greely will not dismiss me." He was dismissed without comment. He went on a drunk, was taken to the hospital where

SECTION DIRECTOR AT GALVESTON

he died. He was buried at Government expense as a veteran of the United States Army.

Another sad case was that of a brilliant young married man who did not drink, but who was an inveterate poker player. Poker playing materially impaired his efficiency. He was often tardy in reporting for duty. One morning, soon after I reached the office, he telephoned an assistant and asked him to carry on his work for a while as he was going fishing. He said that he would be at the office by the time that assistant had to take up a regular detail at 8:30 a. m. A Western Union messenger came into the office at 8:30 with a note from the man stating that he was at home sick and could not come to work. He had reported sick for short periods so often that I decided to check on him. I boarded a street car and went to his house. As I approached the house a colored man was leaving and the man's wife was standing at the gate. She informed me that her husband was very sick and asleep and asked me to please let her telephone me about his condition when he awoke. I boarded the same street car that had brought me to his house and went to the office. When I reached the office there I saw the man, who was supposed to be at home sick in bed, performing his regular duties. He stopped, stunned for a minute, then went on with his work. Not a word was said but the incident cured him—to some extent at least, for he was prompt in the performance of his duties after that occurrence. His wife told me that she impressed on her husband that she would never lie for him again; that being caught outright once was enough.

CHAPTER XI.

SUMMER HOT WINDS ON THE GREAT PLAINS

"WESTWARD, Ho!" was echoing throughout the country in the late 'sixties and through the seventies. The soil of the Great Plains region when the rains are copious blossoms like a paradise. Soon after the close of the War between the States farmers drifted westward to the fringe of civilization. A succession of good seasons made them wealthy in a few years. They made visits to their old homes and spread the news of the paradise they had found. Farmers, merchants, "the baker and the candle-stick maker" packed their belongings in covered wagons and trucked westward to seek their fortunes in the untried land of promise. They never stopped to consider that "every rose has its thorns". They visualized nothing but prosperity and future wealth. Schools and colleges were established and some of the most brilliant men in the country cast their lots in the rapidly developing Great Plains region. These men had vision, they made records of the meteorological conditions in their localities from day to day in cooperation with the United States weather service, which had come into its existence as the western exodus was at its zenith. They expected to find that the rains would come regularly and never dreamed that periods of destructive drouth, attended by hot winds would cause promising crops to be burned until the fields so green and beautiful today would be bleak and brown tomorrow. Little did these co-operative observers dream that soon they would be recording weather features such as the following:

> Lawrence, Kansas, September 12-15, 1882: "Fierce dry blasts burned the foliage of the trees so that they crumbled to powder at the touch of the hand."

> Leavenworth, Kansas: September, 12-14, 1882: "Hot winds withered and burned up vegetation."

Rocky Mountain locusts added further destruction to vegetation, and the country presented a desolate scene.

Frequent and refreshing rains again brought prolific crops. The spirits of the people of that region were revived, and many who had gone back East returned to the plains region to try their fortunes again. However, they were destined to suffer in the future from these hot winds at irregular intervals. The following are a few of the observations of the effects of these winds which were published in the Bulletin of the Philosophical Society of Washington in 1894:

> Brownsville, Nebraska, July 29, 1887: "Brisk southwest winds, hot as the breath of a furnace. Corn and all vegetation burned up. Apples baked on the trees in some instances."
>
> Shaky, Kansas, July 27, 1889: "Hot winds from the north made metals so hot that they were uncomfortable to the touch."
>
> Northern Texas, Many Stations, May 29-30, 1892: "Hot winds traveled in narrow streaks in several counties, scorched young corn, and other vegetation to a crisp."

The unusual features of these hot winds and their damage to crops in Texas, an area under my supervision, impressed me so forcibly that a study of the HOT WINDS OF THE GREAT PLAINS Region was made in the interests of agriculture. My purpose in the study was to determine the conditions which caused them, and to find out if possible what action might be taken to ameliorate their harmful effects. A temporary assignment to the Central Office of the Weather Bureau in Washington afforded me an opportunity to make a thorough study of this subject. All the observation records made at stations in the Plains Region were examined day by day, and extracts made of all references to destructive hot winds. By invitation the study was presented to the members of the Philosophical Society of Washington in an address in which I gave my findings. They published the study in full.[*]

Fifty years have passed since the bulletin was printed and no contributions have given us further enlightenment on this

---

[*] Under the following title: SUMMER HOT WINDS ON THE GREAT PLAINS, U.S.A., Isaac Monroe Cline, Philosophical Society of Washington, Bulletin Vol. XII, pp. 309-348. Plates 4-6, Washington, D. C., 1894.

subject. These drouths and hot winds are the result of cyclonic action. Cyclones that move slowly southward over the Plains Region, or remain nearly stationary for three or four days and sometimes longer in summer, draw masses of North Pacific air across the Rocky Mountains to the low pressure area. In ascending the western slope of the Rocky Mountains, the cooling of this air mass in ascent is attended by condensation and precipitation; the temperature of the ascending air mass is higher, at the elevation where condensation is taking place, than it would have been if no condensation had occurred. The air mass reaches the eastern side of the Rocky Mountains very dry and with a relatively high temperature. We quote from the Bulletin of the Philosophical Society, 1894, mentioned above:

> Eight tenths of the moisture of the atmosphere is below the crest of those mountains, and air in passing over them loses a large percentage of this moisture. This dry air in descending over the eastern slope, after having dissipated the cloud carried over, gains temperature dynamically nearly twice as rapidly, in a corresponding distance, as it cooled in ascending the western slope. In moving towards the low pressure area this dry air takes up the circulation around that area, is carried over the Plains Region from a northerly direction, and flows down over the eastern slope from a **westerly and then** a southerly direction depending on the trend of the isobars. The dry air is carried forward in the upper strata more rapidly than in the layers near the earth's surface and when thus carried out over the moister and less dense air its tendency is to descend here and there through that air to the surface. These currents (or air masses) in descending rapidly from great elevations gain a great deal of warmth and reach the earth with their initial dryness.

Recommendations made in this study fifty years ago, if carried out, would have brought some relief to farm crops when hot winds occur. It was suggested that forests be planted along the streams in the Great Plains as wind brakes. The forests would reduce the geographical extent of these hot winds and their injurious effects on vegetation.

However, periods of bounteous crops and periods of poor crops may be expected to occur through all time. Hon. Henry

C. Wallace, when Secretary of Agriculture, appears to have given very little thought to the fact that we have fat and lean years, resulting from the acts of God, otherwise he would not have ordered the killing off the meat supply and the reduction of crop acreages. He would have displayed wisdom if he had done as Joseph told Pharoah to do, lay up in the fat years so that you will have something to eat in lean years. Millions throughout the world were starving and they could have been fed with our surplus if it had been properly handled instead of being destroyed. We have suffered from Wallace's blunders. He evidently did not go to the Bible for the translation of his dreams or he would have discarded many of them and the world would have been better off by his having done so.

CHAPTER XII.

MY TRANSFER TO THE DEPARTMENT OF AGRICULTURE

WEATHER conditions are so intimately related to agricultural interests that the weather service and its employees were transferred to the Department of Agriculture by Act of Congress on July 1, 1891. General Greely had made the weather service one of the most popular branches of the Government service, and he was retained for a while after the transfer both as Chief Signal Officer and Chief of the newly created Weather Bureau. Prof. Mark W. Harrington, editor and publisher of a meteorological journal, succeeded General Greely as Chief of the Weather Bureau in 1893.

Four hundred years had passed since Christopher Columbus discovered land in this part of the world on October 12, 1492. The anniversary of that important historical event was celebrated by holding a great international fair. The World's Fair at Chicago was organized on a grand scale, and the advancements made in science were not overlooked. A world's Scientific Congress was convened as part of the celebration. Outstanding scientists from the leading nations of the world attended, and they discussed the scientific achievements of their respective countries. The United States Weather Bureau was represented, for the weather service in the country which Columbus discovered was the outstanding weather service of the world. I had the honor of being a member of that distinguished scientific congress.

Weather forecasting was nothing more than a listing of probabilities at that time. But Chief Harrington visualized that weather changes could be forecast with considerable accuracy. In order to get good forecasters he arranged for a competitive examination, the competition being open to any one who desired to enter the contest. Each competitor had to submit a paper on the subject, "Improvement of Weather Forecasts". Some thirty papers were submitted, all under assumed names. (Actual names in sealed envelopes attached were not to be opened until the papers had been examined and graded.) The authors of the ten papers receiving the highest marks were instructed to report to the Chief of the Weather Bureau at Washington, at their own expense, to finish the contest. With a wife and three children to support,* the expense of the journey appeared to eliminate me. My dilemma was explained to a friend, Mr. George Sealy, Vice-President of the Gulf Colorado and Santa Fe Railroad. He told me to go ahead with the contest, and that he would get me passes over the railroads from Galveston to Washington and return. Mr. Sealy furnished me railroad passes and I went to Washington to continue the contest, which lasted from January through March, 1894. This experience had much to do with my future advancement and success. Two men tied for first place, Willis L. Moore and William H. Hammon, and both were made forecasters. I was fifth with a grade of only three-tenths of one per cent behind the winners. A warm personal friendship was formed with Willis L. Moore, which in later years was to become disrupted.

I was instructor in Medical Climatology in the Medical Department of the University of Texas, and studies of weather changes as they affected persons suffering from various diseases were carried forward during my recreation time. I also matriculated at Add Ran (now Texas Christian University), studied philosophy and sociology under Dr. J. W. Lowber, President of the University, passed the examinations and received my Ph.D. degree. This course improved my knowledge of English Literature and has been most helpful in my work.

Prof. Willis L. Moore succeeded Prof. Mark W. Harrington as Chief of the Weather Bureau in 1895. Prof. Moore displayed great ability and followed the policy instituted by General Greely. He continued the elimination of men who were

---

* My second child, Rosemary, was born on August 24, 1889, and my third child, Esther Ballew, on July 2, 1894.

not of good character. A telegram of one hundred words enumerating a list of charges against the forecaster at one of the most important stations in the country was received late one afternoon. I was directed to proceed immediately to that station and investigate and report on the unbecoming conduct of the official in charge. I took the next train, and when I arrived at the city in question the first thing I saw was newspapers with blazing streamer headlines telling about the scandalous conduct and disappearance of the weather forecaster. This man was a beau brummel who did not hesitate to overstep the bounds of propriety. He had apartments handsomely furnished on the third floor of an important office building where his visitors could more easily escape suspicion. However, "murder will out". I soon found a source that furnished much information about his amours. A publication, which appears to have had for its object the collection of information concerning the secret escapades of men and women, was brought to my attention. The publisher, when informed that I was a Government official making an authorized investigation for the Weather Bureau, placed all the information he had at my disposal. There were several galleys of type describing the amours of this forecaster which had not been printed, because he had evidently "paid off". This unpublished matter was rich, rare and racy. One article described how he filled the bath tub with Florida water, and bathed his lady visitors. Our Lothario evidently entertained demi-mondes in his apartment, or his activities would not have come to the attention of the editor of the paper. One of the unpublished articles told of an instance where an irate parent went to the forecaster's apartment looking for his daughter. The forecaster saw the father coming up the stairs, took the girl out through a window and down a fire escape before her father got into the apartment. However, the particular case which I had come to investigate was that in which a man had been tipped off by some one in the building that his wife was frequenting the forecaster's apartment. The husband arranged with a watcher to notify him the next time she went there. He was notified and appeared at the apartment prepared to kill. He found the door locked. Demanding admittance and getting no response, he then proceeded to break the door down. When he got into the apartment, it was empty. There was not a sign of suspicious evidence; apparently the Florida water stage had not been reached. There was a large wardrobe in the apartment, and while the irate husband was breaking in the door, the forecaster picked up the woman who was small and placed her on the top shelf of the wardrobe, locked its door and put the key on the sill of the open window. He then jumped out of the window and escaped over the roofs of the adjoining buildings, making his get-away before the husband got inside the apartment. The husband was certain that his wife had been in the apartment and suspected that she was locked in the wardrobe. He shot the wardrobe full of holes, but not higher than a person's head. I imagine that the woman on the shelf of the wardrobe fainted while the bullets were flying under her. The forecaster in the meantime telephoned a friend who was acquainted with the apartment, asking him to go there, get the key off the window sill, unlock the door of the wardrobe and take the woman, if still alive, to the street and let her go her way. The friend told me that he found the woman safe on the top shelf of the wardrobe, but that six bullets fired into the wardrobe had frightened her out of her senses. He gave her some aromatic spirits, accompanied her to the street where she left presumably to go home to meet her angry husband. This story was given me under oath in writing. The weather forecaster's hat and coat were found on the lake shore and it was assumed that he was so badly frightened that he had drowned himself. I came across an unused newspaper artists' cartoon inspired by the incident. The scene was hell; his Satanic Majesty was seated on a throne of red hot steel, and there in the midst of the flaming fires was the late forecaster with a piece of charcoal writing on an asbestos board, "A cold wave for Hell and vicinity". The cartoonist had added satirically underneath, "And this is just as likely to be verified as the forecasts he made for this locality". The last heard of the forecaster he was speeding towards the Pacific coast, and the irate husband with two pistols in holsters buckled around his waist not far behind him. The forecaster was summarily dismissed from the Weather Bureau.

Soon after the above incident I had occasion to investigate the conduct of another forecaster at an important station. This official had caused the separation of a man and his wife under such circumstances that the scandal reached the newspapers. The woman was described as being as beautiful as Venus. In making the investigation I went to the residence of the husband to get a statement from him. I rang the door bell and who do you think answered it? It was the beautiful woman! Her husband was sick and had taken her back. I recognized her from the descriptions given me. I said: "I beg your pardon, this is the wrong number". The facts were reported and the official was dismissed.

CHAPTER XIII.

COOPERATION OF MEXICAN WEATHER SERVICE WITH U. S. WEATHER BUREAU.

OUR country was evidently preparing for war before the battle of Manila. Plans were being made before that occurrence to extend the range of weather observations so that warnings of tropical cyclones attended by hurricane winds which move up through the Caribbean Sea, the Gulf of Mexico and the Atlantic Ocean, could be given to ships at sea. This extension of the warning service was primarily, though not so specifically stated, for the guidance and protection of our fleet of war vessels in the Caribbean Sea. Every effort was being made to keep our warships informed concerning the movements of cyclones so that they could move out of their paths and escape danger. Sometime before the commencement of hostilities, instructions were received at Galveston, directing me to proceed to the City of Mexico and confer with the Mexican Government Officials with a view to securing the cooperation of that Government in the collection of weather observations on the Mexican Gulf coast and in Yucatan. Standard mercurial barometers were taken with me and comparative observations were made with the standard barometers of the Mexican Meteorological Service in the City of Mexico. A weather service was being organized in Mexico under the management of the Director General of Federal Telegraphs of Mexico. The Mexican Government gave the United States Weather Bureau permission to establish weather observation stations at Tampico, Vera Cruz, and Coatzacoalcos, on the Mexican Gulf coast, and at Merida and Progresso, Yucatan. Managers of the Mexican Cable Company, an English concern, served as observers on the Gulf coast of Mexico. The cable company transmitted the observations from their stations to the Weather Bureau at Galveston without expense to the United States. The observations from Progresso and Merida were forwarded to Galveston through the cooperation of the Federal Telegraphs of Mexico. The instruments used were not efficient or up-to-date, and the observers were informed that better instrumental equipment would be furnished them shortly.

New instruments for the above stations were awaiting me at Galveston when I returned. I was directed to proceed immediately with the installation of the new instrumental equipment, and to instruct the observers to commence observations as soon as the instruments were placed in position. Tampico was the first station equipped. Then I visited the Mexican Officials in Mexico City to make arrangements for my trip with the instruments to Yucatan. The journey from Mexico City to Vera Cruz was by train. There were no dining cars, but the train stopped for the crew to rest in the middle of the day. The conductor pointed out a place where I could get something to eat. I had enough knowledge of Spanish to make myself understood, especially when something to eat was concerned. When I told the proprietor of the hotel that I wanted a quart cup of coffee, half milk, he went to the door and called. Immediately a fine female goat put in her appearance. He milked the goat in my presence and brought me the milk. That coffee was so good that I can almost taste it now. A cousin of my mother had a goat ranch, and made cheese. I was very fond of both goat milk and cheese.

Commodore George Dewey, U. S. Navy, was assigned to the command of the Asiatic Squadron about the time I made the first trip to Mexico—just a coincidence. When I arrived at Vera Cruz, on my second trip, I learned that the Asiatic Squadron under command of Dewey had annihilated, in the battle of Manila Bay, on May 1, 1898, the Spanish Asiatic Squadron, destroying eleven and capturing all other ships as well as all land batteries without the loss of a single man. When Dewey attacked the Spanish Fleet the German fleet started to the assistance of the Spaniards, but the Commander of the English Fleet trained his guns and signaled the German Commander that if he made a move the German fleet would be blown to bits. The Germans remained neutral. Vera Cruz was boiling over with excitement. Sentiment was with the Americans.

Mexican Officials had told me that the journey from Coatzacoalcos to Progresso would be on a small coastal steamer which had recently been transferred from the Spanish to the Mexican flag. They gave me a letter to the captain of the boat; the letter did not mention my nationality and while I thought this strange I asked no questions. The Cable Office at Coatzacoalcos was in charge of an English East Indian. He informed me that the captain of the boat, on which I would have to travel along the coast for seven days, was very bitter against the Americans. He said that my nationality was omitted from the letter intentionally, and that working with the cable offices the captain of the boat would assume that I was an Englishman. The previous trip to Progresso had been made on a Ward liner. The manager of the cable office introduced me to the captain of the boat and I was immediately assigned to a State Room or Cabin. At meal time I was placed at the right hand of the captain. The Gringoes, as we were called, were cursed continuously, which emphasized the bitterness of the captain and crew who were all Spaniads. I spent most of the time in the cabin and kept quiet. If they had known for a fact that I was a Gringo, I might accidentally have fallen overboard and become a meal for the sharks.

Progresso and Merida, Yucatan, were having the worst yellow fever epidemic they had ever experienced. The yellow fever did not deter me from my duty for I went there to help win the war. The commander who knows his weather will win when another who does not know his weather would lose the battle. I was there to help get weather information to the commander of our fleet of war vessels in the Caribbean Sea. I remained in Yucatan for some time and hundreds of deaths occurred daily in that part of the country. My knowledge of medicine stood me well in hand. I had already espoused the theory that the mosquito is the carrier of the dreaded disease and that it communicated it to the human being. I occupied a screened room in the United States Consulate and slept in a hammock. I went to my room before the sun went down and remained there in the morning until the sun was up. The yellow fever did not attack me. Dr. Walter Reed and other investigators were carrying on experiments in Cuba, at this time, to find out whether or not the mosquito is the carrier of yellow fever. Lives were sacrificed but not in vain. Dr. Reed and his co-workers met with much opposition, as is the case with every investigator who advocates a new theory. Dr. Reed proved beyond question that the mosquito is the carrier of that death-dealing disease, called "Yellow Jack". His work enabled others to prevent the terrible epidemics which had in the past claimed so many lives. Dr. William C. Gorgas followed Dr. Reed, and applied methods for combatting and eliminating the mosquito carrier, so that the disease is no longer such a menace. The service rendered to humanity by these men is so great that no value can be placed upon it.

Food was scarce in Yucatan, for the blockade was quite effective. Ships were afraid to sail in that region. When a lobster could be found I would buy one and get the restaurant to cook it for me.

Richard P. Hobson, who with a crew of seven other volunteers sunk the collier Merrimac in Santiago de Cuba harbor, were caputred and held prisoners in the Spanish fortress, from June 3rd to July 6th, 1898. These men were willing to sacrifice their lives for their country.

Commodore Winfield S. Schley, of Arctic fame, commander of the flying squadron, was in immediate command of our combined fleet at the battle of Santiago de Cuba, on July 3, 1898. The Spanish Admiral Cervera's fleet was completely destroyed. This information reached me at Progresso on the date of the battle.

Success had crowned my efforts in the work I was doing to help in winning the war. I was now ready to return to my station at Galveston. A Ward line steamer was passing Progresso about the middle of July, and I secured passage to Vera Cruz. When I boarded the ship the surface of the water was smooth and quiet. An occasional long swell would lift the steamer up and let it down slowly. A brickdust sky portended

COOPERATION OF MEXICAN WEATHER SERVICE 83

stormy weather. The dining room was crowded with one hundred people for breakfast, all in a jolly mood. The sky darkened, torrents of rain came down, the wind roared and increased in force until the steamer rocked and pitched with the hurricane winds like a bucking broncho. When lunch time came I was the only person who showed up in the dining room, the dread sea sickness had not yet upset me. When time for the evening meal arrived I was so sick that I could not leave my stateroom. I called the steward and asked for an orange which I thought might settle my stomach. I was so sick that I did not care if the ship went to the bottom of the Bay of Campeche. This was my first experience in a tropical cyclone, but it was not to be my last.

Additional reports from weather observing stations south of the United States were needed to improve weather forecasting. An exchange of weather reports between the United States Weather Bureau and the Mexican Weather Service was recommended. I returned to Mexico City clothed with authority to complete arrangements for such an exchange. The Director General of Federal Telegraphs had supervision of the telegraphic weather service in Mexico, and the observations were taken by the telegraph operators at the different telegraph stations. The Superintendent of the Mexican Cable Company told me that the Federal Telegraphs of Mexico had a contract with the Western Union Telegraph Company whereby the Western Union Telegraph Company was to transmit weather reports free of charge for the Federal Telegraphs to and from any points designated by the Director General. This enabled me to devise a plan of exchange equable to both Mexico and the United States. I proposed to the Director General of the Federal Telegraphs that they deliver their weather reports to a station in the United States designated by me without expense to the United States Weather Bureau, and that we would collect the reports from such stations as he desired and file the message with the Western Union Telegraph Company for transmission to the City of Mexico without expense to the United States. We could give him twice as many reports as he could give us and the plan of exchange was based on those grounds. The Director General approved the plan, the exchange to take place at Galveston, or such other point I might suggest in the future. The exchange went along without a hitch for quite a while and then the telegrams came in collect form Eagle Pass over Western Union lines. I telegraphed that fact to the Director General of Federal Telegraphs, Mexico City, and he had the Western Union cancel the charges and resume sending them to Galveston DH and DH. On another occasion, some years later, the accounting office of the Western Union Telegraph Company in New York filed a bill with the Telegraph Division of the U. S. Weather Bureau at Washington for several hundred dollars back tolls for transmission of these weather reports over Western Union wires. The officials in Washington were going to pay this bill over my protest. When I learned that Washington was determined to pay the Western Union the amount they claimed I telegraphed the Director General of Federal Telegraphs, Mexico City, and laid the facts before him. The Western Union withdrew their bill from the Weather Bureau Office in Washington the following day. The exchange service then went on smoothly up to the time of my retirement from the Weather Bureau.

CHAPTER XIV.

TEMPERATURE FORECASTS—
THE BLIZZARD OF FEBRUARY, 1899.

IN the vicinity of Galveston with its almost tropical climate, the growing of sugar cane and winter-grown vegetable truck for the markets was profitable. While freezing temperatures occurred on an average of about five times a year, a freeze sometimes caught the crops before harvesting had been completed, and heavy losses resulted. Soon after my arrival in Galveston the growers asked me to give them warnings when freezing temperatures were indicated, and to predict what the temperature would be in the following 24 to 36 hours. My predecessor had turned down such requests. Growers stated that with this information they could protect the crops still in the fields and prevent much of the loss they had been sustaining from freezes. They were informed that such forecasts could be made with success, and that I could give them the service they desired, but that authority to issue such forecasts would have to be obtained from the Chief of the Weather Bureau in Washington. The representations of the sugar and truck growers were laid before the officials in Washington in 1894. My clients wanted me to go ahead pending the approval of the authorities at Washington, but I made it a rule to obey orders unless the public interest was in immediate danger. An official who did an excellent piece of work without specific authority was likely to be looking for another job as shown by an experience I will report on in another chapter. The Chief of the Bureau wrote me that the scientific staff had looked carefully into the matter and reached the conclusion that it would be impossible for any one to make a forecast of what the temperature would be 24 to 36 hours from the time the forecast was issued with any degree of accuracy. Surrender never won a battle, and battles that might have been won are sometimes lost in surrender. My motto has always been, "fight to the last ditch". Practice forecasts were made, on my own initiative, for the localities involved during the winter of 1894-1895. These practice forecasts showed that temperatures could be forecast with a marked degree of accuracy. Willis L. Moore, one of the leaders in the forecasting contest mentioned in a previous chapter, had been appointed Chief of the Weather Bureau in 1895, while I was making the practice forecasts. My practice forecasts were submitted to him in full, showing their percentage of verification. Prof. Moore did not consult the scientific staff, but personally authorized me to give the public the temperature forecasts they desired. This was making history in weather forecasting. These were the first forecasts to be issued, warning the public what temperature to expect for the succeeding 24 or 36 hours. These forecasts were so successful and of such great value that soon afterwards every forecaster was directed to issue such predictions for his locality. Such forecasts not only enabled growers to protect crops, and cattle growers to protect livestock from severe cold, but house owners to drain exposed water pipes and motorists to drain radiators to keep them from bursting. The uses now made of these forecasts are too numerous to mention. Temperature forecasts enabled Texas sugar and truck growers to save crops worth thousands of dollars annually, which otherwise would have been a loss. The importance of these forecasts can be realized from the fact that when in 1901 the Forecast Center was removed from Galveston to New Orleans, and I was transferred with it, Col. Ed H. Cunningham of Sugar Land, Texas, representing the sugar interests of that region, wrote the Secretary of Agriculture relative to the special temperature forecast, in which he stated that my transfer from Galveston would result in disaster to the sugar interests of Texas. The letter was referred to me at New Orleans, and arrangements were made so that I could give the sugar and trucking interests of both Texas and Louisiana the same complete service I had rendered from Galveston.

Blizzards which frequent the eastern Rocky Mountain Slope seldom reach the Gulf Coast. During my seventeen years of service only one severe blizzard ever forced itself through the semi-tropical heat wall which protects the Gulf Coast. Blizzards are attended by high winds and low temperatures over the eastern Rocky Mountain Region, but are nearly always toned down to moderate conditions before they reach the Gulf Coast. During the early days of February, 1899, a mass of cold air with unusually high barometer and low temperature remained nearly stationary for several days over the eastern Rocky Mountain Slope. This mass of cold air was gathering force not visible to the intuition of man, and on February 10th, it commenced moving southward with such great rapidity that it caused a blizzard over the entire South. The temperature at Galveston was 8 degrees on February the 12th. The blizzard rode on the wings of winds ranging from 40 to 80 miles per hour—hurricane winds from the north. The winds blew the water out of Galveston Bay, except in the ship channels, and the water in these moved with the rapidity of mill tails. The dry bottom of the Bay was covered with wild geese and other fowl. Two of us in a four-oared Saint Lawrence River skiff attempted to cross a channel 100 yards wide and shoot geese, but before we could get our oars to working we found ourselves carried one mile towards the sea without making any headway across the channel. Escape appeared impossible, but we had the good fortune to run aground on a point of shoal jetting out from the Galveston side of the Bay. When we struck that shoal we jumped out and pulled the skiff ashore. We let the geese stay where they were. However, when the wind subsided we got the geese. This calls to mind a story which I will repeat here:

> A man 70 years old married a girl of about 20 summers. The union resulted in a baby boy. When the boy was about one year old the girl having grown tired of the old man ran off with a younger man. The old man loved her devotedly and her loss grieved him and caused him much suffering. He resolved that his boy should never suffer as he had suffered on account of a woman. He put the boy in a boarding school far from civilization where he thought no woman would venture. The old man visited the boy every Christmas and they would take long walks. The boy was now fourteen and the old man was 86 years old. The boy had never seen a woman. A school for

girls had just opened about five miles from the boys' school, but the boys had not yet learned of this. The old man took the boy over a route they had never taken before. This route passed the girl's school and the girls were romping on the campus. The old man was walking with his head bowed and still grieving for his young wife, when the boy suddenly grabbed him and said: "Dad, look, what are those?" The old man looked up and said: "Son, those are geese." The boy looked at them for a moment and said: "Dad, I want one of those geese."

The Trinity River, discharging its icy waters into Galveston Bay, had chilled the waters of the Bay to near freezing and brought ice floes down into the channels. The water in the channels of the Bay was so cold that it chilled the fish till they could not swim and they floated up to the wharf where they could be picked up by hand. The temperature was 6 degrees below zero at Abilene, Texas, and 5 degrees below at Shreveport, Louisiana. New Orleans had a minimum temperature of 6.8 degrees on February 13th. The severe cold caused the death of some of the maskers in a carnival parade. Great ice gorges formed in the Mississippi River above New Orleans and floes of ice with icebergs more than ten feet in height passed down the river by New Orleans for more than two weeks.

CHAPTER XV.

FLOODS IN THE COLORADO AND BRAZOS RIVERS.

DAMAGE by floods in the low lands along Texas streams occurs occasionally, while along stretches of the same streams there are times when irrigation is necessary to grow crops. Precipitation records made under my supervision, as Director of the Texas Section of the United States Weather Service, were studied carefully both for advising as to the amount of water available for irrigation of growing crops, and for use in the issue of warnings when floods were indicated. The amount of run-off from rains of different intensity was worked out for the drainage area of each important stream; then the volume of water carried by the stream, as it crossed each degree of latitude on its way to the Gulf, was determined. This information ready for quick reference enabled me to act promptly in giving advice on the probability of a flood as well as warnings to the public.

Flood warnings services had been authorized by the Chief of the Weather Bureau for some of the larger rivers. However, all warnings had to be submitted by telegraph to the Central Office of the Weather Bureau at Washington, for approval before they could be released to the public.

Heavy rains early in April, 1900, had brought the larger rivers in Texas almost to danger stages, and it appeared that destructive floods were in the making. Hours of duty were never counted when there was something to be done for the welfare of the public. Midnight often found me working on flood problems. Floods damaged property and often took life, and I stayed on the job twenty-four hours at a stretch when necessary. Such work never tired me for I realized that I was helping some one, somewhere.

Flood conditions became ominous, and something was to happen along with these rains which the ken of human minds could not foresee. Preparedness enables us to act when things out of the ordinary occur. Heavy rains fell over the drainage basin of the Colorado River on April 7, 1900. The run-off from this precipitation had to be determined without delay, and everything made ready for any emergency. The already swollen river would reach a dangerous stage if heavy rains should continue one more day. On Saturday conditions were so threatening that I waited for the night reports. About 6:30 p. m. a telegram was received from the co-operative observer at Austin, which said: "Six and one-half inches of rainfall, dam broken". This meant that a flood of great proportions would sweep down the Colorado River with unusual swiftness. The great reservoir dam across the Colorado River at Austin was built of granite and was considered strong enough to withstand any flood. The dam was 1,250 feet long and 65 feet high. The width of the water in the reservoir was 1,000 feet and 65 feet deep, and the lake created by the dam extended upstream thirty miles. This tremendous volume of water with a 65 foot head, pouring suddenly upon the flood water already in the Colorado bottoms threatened the destruction of property of great value and placed thousands of human lives in jeopardy. Strict regulations required that emergency warnings of any nature be submitted to the Chief of the Bureau in Washington for approval before being distributed to the public. It was then 6:30 p. m. Telegraph offices in the low lands along the river would soon close for the night. The following day was Sunday, and many of the telegraph offices would not open until Monday. There was no time to communicate with Washington and get approval for emergency flood warnings. I neither had authority to issue emergency warnings nor to incur the telegraphic expense involved. But red tape had to be cut, for delay meant the loss of property and human lives. Telegrams were sent immediately to every postmaster in the Colorado Valley, between Austin and the Gulf of Mexico, telling them that the dam had washed out and that a tremendous volume of water was moving rapidly down the valley. All persons residing in the low lands along the river were warned to move out with their belongings. The advices were obeyed. People, with their belongings, moved to places of safety and not a life was lost in the great and sudden flood, except that two persons were drowned where the dam gave way. The telegraph and the telephone enabled me to render this service.

The same spring heavy rains fell over northern and central Texas the latter part of April, 1900. The calculated runoff from these rains indicated unprecedented floods, which would overflow the bottoms of the Brazos River during the early part of May. Flood warnings, ten days in advance of the expected flood, were prepared and telegraphed to the Chief of the Bureau at Washington, and authority was received to telegraph the warnings to the public. The greatest flood in the history of that river overflowed the bottoms. People in dangerous locations moved out with their belongings so that no loss of life resulted. Growing crops, however, suffered much damage.

CHAPTER XVI.

CYCLONE OF SEPTEMBER 5-10, 1900—
THE GALVESTON HURRICANE.

HISTORY does not record a greater disaster in the United States, than that which occurred at Galveston, Texas on September 8, 1900. A tropical cyclone made its appearance early in September and was north of Cuba on the 3rd, passed through the Florida Straits on the 4th and 5th and traveled in a northwesterly direction through the Gulf of Mexico. It was centered about 200 miles south of the Mississippi River at 8 a. m. on September 7th with the barometer reading 29.65 inches and the wind 30 miles per hour from the northeast at New Orleans. At the same hour the barometer was 29.70 inches with a north offshore wind of 10 to 13 miles per hour at Galveston, Texas, the storm center at that time being about 400 miles from Galveston. There were no ship reports in those days and New Orleans was the nearest station of observation to the center of the cyclone. The cyclone was advancing towards the Texas coast with a speed of about 12 miles per hour while the winds near the center of the cyclone were of hurricane force and probably exceeded 100 miles per hour. The barometer at Galveston changed very little during the 7th and up to 2 a. m. of the 8th.

I had been in the weather service some 18 years, and 11 years in Galveston, but during this time the only severe tropical cyclone that had reached the Gulf Coast was that of September 30 and October 1, 1893. This cyclone destroyed Chenier Camanada and drowned 2,000 persons on the Louisiana, Mississippi, and Alabama coasts. A tropical cyclone on August 9, 1856, drowned 300 persons who were summering on Last Island, Louisiana. I had studied the meagre information available relative to tropical cyclones. I had read of the Calcutta cyclone on October 5, 1864, which caused a storm tide 16 feet deep over the delta of the Ganges and drowned 40,000 persons, and the Backergunge cyclone of October 31, 1876, which caused an unprecedented storm tide ranging in depth from 10 feet to nearly 50 feet over the eastern edge of the delta of the Ganges, and drowned at the lowest estimates 100,000 persons. The enormous loss of life by drowning in these cyclones had impressed me, and I realized that if this cyclone moved in over Galveston great destruction would result.

September 7th was an ideal day as far as weather was concerned, a calm before the storm, but long swells broke on the beach with ominous roaring and these were building up a tide above the average height. I was up making observations at 5 a. m. September 8th, and found the Gulf water coming in over the low parts of the city with the tide 4.5 feet, on the automatic gauge, above what the tide should have been. The barometer was falling slowly, the wind was 15 to 17 miles per hour from the north, offshore, and the tide was rising with this wind directly against it; such winds under ordinary circumstances cause the tide to fall and give a low tide. The storm swells were increasing in magnitude and frequency and were building up a storm tide which told me as plainly as though it was a written message that great danger was approaching. Neither the barometer, nor the winds were telling me, but the storm tide was telling me to warn the people of the danger approaching.

Complete observations, describing the weather and unusual tide conditions, were telegraphed to the Central Office of the Weather Bureau, Washington, every two hours from 5 a. m. until all the wires went down. Early on the morning of September 8th, I harnessed my horse to a two wheeled cart, which I used for hunting, and drove along the beach from one end of the town to the other. I warned the people that great danger threatened them, and advised some 6,000 persons, from the interior of the State, who were summering along the beach to go home immediately. I warned persons residing within three blocks of the beach to move to the higher portions of the city, that their houses would be undermined by the ebb and flow of the increasing storm tide and would be washed away. Summer visitors went home, and residents moved out in accordance with the advice given them. Some 6,000 lives were saved by my advice and warnings.

The swells sent out by the cyclone continued to increase in magnitude and frequency during the 8th, and the storm tide rose rapidly. At 3 p. m. on the 8th the storm tide was 8 feet on the automatic gauge and the entire Island was covered with water from the Gulf to the Bay. The wind continued offshore against the rising tide until 4 p. m. when it shifted to the northeast. The barometer fell 1.00 inch from 29.48 at 12 noon to 28.48 at 8:30 p. m. of the 8th, when the center of the cyclone passed over the city. The wind had been north, except northeast at irregular intervals, up to and including 3 p. m.—northeast 4 p. m. to 8 p. m., east 9 and 10 p. m., southeast 11 p. m. and south at midnight. After 3 p. m. of the 8th, when the wind shifted to northeast and east the storm tide rose with phenomenal rapidity, reaching 15 feet at 8 to 9 p. m., a rise of 1.4 feet per hour. From 6 p. m. to 8:30 p. m. the storm tide rose at a uniform rate of about 2.5 feet per hour. There was no sudden rise in the water with the passage of the center of the cyclone. The tide fell rapidly after the passage of the center of the cyclone and was below 7 feet at midnight. There was no sudden, extraordinary rise in the water with the passage of the center of the cyclone, and claims made to that effect by some in after years are incorrect. (For exact height of the water see tide graph, page 244, TROPICAL CYCLONES, Cline, The Macmillan Company, New York).

I recognized at 3:30 p. m. that an awful disaster was upon us. I wrote a message to send to the Chief of the Weather Bureau at Washington, advising him of the terrible situation, and stated, that the city was fast going under water, that great loss of life must result, and stressed the need for relief. I gave this message to my assistant, Dr. Joseph L. Cline, who wading through water nearly to his waist, carried it to the telegraph office but found that all telegraph wires were down. He then went to the telephone exchange and found that they had one wire to Houston which was working intermittently. The telephone company turned this line over to him, and after repeated efforts, he got the message through to the Western Union telegraph office in Houston just as this line went completely out, cutting Galveston off from the outside world, and leaving the inhabitants of the city to the fury of the winds and storm tides, with no means of telling the world of the catastrophe which the storm tides and hurricane winds held in store for the people that awful night.

As the telephone and telegraph wires were all down, and I could give no further warning nor advices, my services were terminated by the elements. There remained nothing else I could do to help the people. My wife and three little girls were in our home surrounded by the rapidly rising water, which already covered the Island from the Gulf to the Bay. In reality there was no Island, just the ocean with houses standing out of the waves which rolled between them. Having worked for the public ceaselessly from 5 a. m. until 3:30 p. m. and with no possibility of being able to further serve the populace I waded nearly two miles to my home through water, often above my waist. Hurricane winds were driving timbers and slates through the air every where around me, splitting the paling and weather boarding of houses into splinters, and roofs of buildings were flying through the air. My house had been built recently structurally designed to withstand hurricane winds.

After a journey through horrors which were to be worse, I reached my home. Some fifty persons had sought safety therein, among whom were the builders of the house and their families. The rapid rise in the storm tide soon forced us to the second story. Here we could see wreckage being tossed by the winds and waves, driven against buildings and breaking them to pieces, thus increasing the wreckage for the waves to use as battering rams as they rushed forward. Some of this wreckage hung around our house and formed a dam against which the water banked up to a depth of twenty feet above the ground or nearly ten feet higher than the storm tide. We probably would have weathered the storm, but a trestle, about one fourth of a mile long, which the street railway had built out over the Gulf was torn from its moorings, and the rails held it together as one long piece of wreckage. The storm swells were pounding the other wreckage against our home. It held firm against these without trembling. But the street railway trestle was carried squarely against the side of the house. The breaking swells drove this wreckage against the house like a huge battering ram; the house creaked and was carried over into the surging waters and torn to pieces. As it went down my brother, who had come to the house after sending the last message out of the city, stood with my two oldest children near a window which had one large glass, and as the house commenced giving way, he knocked the glass out of the window and took the two children out onto the piece of trestle that was pounding the house to pieces. I with my wife and baby, six years old, were in the center of the room. We were thrown by the impact into a triple chimney and were carried down under the wreckage to the bottom of the water. My wife's clothing was entangled in the wreckage and she never rose from the water. I was pinned under the timbers and thought I would be drowned. The last thought that I remembered that passed through my mind was this—"I have done all that could have been done is this disaster, the world will know that I did my duty to the last, it is useless to fight for life, I will let the water enter my lungs and pass on". However, it was not my time to go. When I regained consciousness, I was floating with my body hanging between heavy timbers which had pressed the water out of my lungs. A flash of lightning revealed my baby girl floating on wreckage a few feet away. I struggled out of the timbers and reached her. A few minutes later during another flash I saw my brother and my other two children clinging to wreckage. I took my baby and joined them on the floating debris. Strange as it may seem these children displayed no sign of fear, and we in the shadow of death did not realize what fear meant. Our only thought was how to win in this disaster.

The battle for our lives, against the elements and the terrific hurricane winds and storm tossed wreckage lasted from 8 p. m. until near midnight. This struggle to live continued through one of the darkest of nights with only an occasional flash of lightning which revealed the terrible carnage about us.

In order to avoid being killed by flying timbers, we placed the children in front of us, turned our backs to the winds and held planks, taken from the floating wreckage, to our backs to distribute and lighten the blows which the wind driven debris was showering upon us continually. Sometimes the blows of the debris were so strong that we would be knocked several feet into the surging waters, when we would fight our way back to the children and continue the struggle to survive. While being carried forward by the winds and surging waters, through the darkness and terrific downpour of rain we could hear houses crashing under the impact of the wreckage hurled forward by the winds and storm tide, but this did not blot out the screams of the injured and dying. Well along in the night a flash of lightning revealed a child, about four years old, floating on some wreckage with no one else near. We succeeded in rescuing her, and placed her with my children. The battle of the storm lasted without interruption from 8:30 p. m. until 11:30 p. m., when the storm tide had receded so that the wreckage we were on touched ground at 28th Street and Avenue P. We spent the remainder of the night in a home near where we landed.

We arranged with the people who resided in the house to keep the child we had saved, promising that we would see that she was provided for. We learned from her that she and her mother were visiting her grandparents, and what had become of them the child did not know. We made a record of the child's name, hoping that we might find some of her family. About two weeks later when people, other than relief workers from the interior, were permitted to come into Galveston, my brother was in a drug store when a man in great distress came into the store and remarked that his wife and their baby were visiting his wife's mother and he feared that he never would see them again. My brother asked him his name and learned that he was the father of the little child we had saved. The joyful father went immediately to get her.

Sunday, September 9th, came with a clear sky, a brilliant sunrise, almost a calm and quiet sea with low tide—a most beautiful day. But oh, the horrible sights that greeted our eyes! The dead scattered through and hanging out of the wreckage, drowned trying to escape or killed by flying timbers, were to be seen by the thousands—the babe in its mother's arms, riches and poverty side by side with garments torn to shreds, men and women who in health and strength only 24 hours before had looked forward to the future with life's fondest hopes, now dead, along with those who had reached four score years and ten—all classes and all ages. The storm tide and hurricane winds had played no favorites. Some weeks later the body of my wife was found at 28th Street and Avenue P, under the wreckage on which we had made the fight for life and won. Even in death she had traveled with us and near us during the storm. Of the 50 persons in my house only a very few survived.

Under orders then strictly enforced no one except the forecaster in the Central Office of the Weather Bureau at Washington, had authority to issue hurricane and emergency warnings. Although the forecaster at the Central Office was kept fully advised as stated above, neither emergency nor hurricane warnings for this disaster were received from the forecaster in the Washington office. Early on the morning of the 8th the rising tide against opposing winds convinced me that great danger threatened Galveston. My studies of the destructive storm tides of the Orient caused by tropical cyclones had given me a working knowledge of what storm tides indicated, and the destruction which they brought in their wake. On my own responsibility, and against strict orders which required that every thing be submitted to the Central Office before being promulgated, I assumed the authority in the emergency and warned the people of Galveston of their danger and advised them what action was necessary to protect their lives and property. If I had taken time on the morning of the 8th to ask for approval from the forecaster in Washington and waited for his reply the people could not have been warned of the disaster. He had already been furnished complete information, and neither hurricane nor emergency warnings were **received**.

THE GALVESTON HURRICANE

The warnings and advices which I gave personally by telephone and telegraph, were so convincing, and were heeded so generally that the loss of life, great as it was, would have been twice as great without them, as later the storm tide became so high that escape was impossible. By noon the highway and railroad bridges connecting the Island with the main land had gone down, and all means of escape from the city were cut off.

As an illustration of the powerful force of the storm tide and winds combined, a large ocean steamer was torn loose from its moorings and carried several miles inland. A canal had to be dug in order to get the steamer back to deep water.

This being my first experience in a tropical cyclone I did not foresee the magnitude of the damage which it would do. Instead of washing away three blocks near the beach as I had warned early in the morning, an area more than six blocks wide was completely washed away, and some 6,000 persons lost their lives.

Some days after the disaster the local representative of the Associated Press came to me with a telegram from his President in New York asking him if hurricane warnings had been displayed. I told him that storm warnings were displayed for two days previous to the hurricane, and that he knew of the urgent warnings and advices I had given the people and how I had gone among them early on the morning of the 8th and told them of their danger. I then asked him if he thought any thing more could have been done to warn the people, and he replied "Nothing more could have been done than was done". I then said, "Telegraph that to your President", which he did and that closed the inquiry without involving the forecaster at the Central Office from whom no hurricane warnings were received.

An editorial in the New York Evening Sun of September 20, 1900, read:

> The warnings which were sent out by Dr. Cline are said to have saved thousands of lives along the coast. The Texas papers show that in some towns and villages and at many plantations and

farms the force of the wind and the rise of the water were necessarily fatal to life as at Galveston, but the inhabitants and residents profited by their information and escaped inland. (See U. S. Monthly Weather Review, September, 1900, pages 371-372.)

A COMMENT ON WORK PERFORMED

Professor Willis L. Moore, Chief of the Weather Bureau at the time of the Galveston hurricane of September 8, 1900, in an article "I AM THINKING OF HURRICANES" which appeared in the American Mercury, September, 1927, (Vol. XII, No. 45) wrote:

> The press dispatches at the time of the Galveston storm stated that Dr. Isaac M. Cline, the head observer of the Weather Bureau there, was one of the heroic spirits of that awful Saturday; that in addition to warning the people by telegraph and telephone, he worked personally among those on the beach on Saturday morning, and long before the waters rolled over the city was driving the people from their homes to higher ground; that when the last means of communication with the outer world had failed, instead of going to the relief of his own home, he braved the wind and the raging waters and reached a telephone station leading to the main land just before the cable parted. He succeeded in sending to me the last message from the doomed city that was transmitted for many days. In this message he stressed the need for relief and said the city was rapidly going under water, and that great destruction of life must ensue.
>
> (Note: I wrote the message mentioned by Prof. Moore, gave it to Dr. Joseph L. Cline, my chief clerk, who performed the feat of getting it filed. I.M.C.)

No written words can describe the horrible conditions that existed in Galveston following the destruction wrought by the storm tide and hurricane winds. Martial law was declared to protect the living from thieves and the dead from ghouls. The low criminal element, both white and colored, would cut the heads off the dead to get their necklaces and the fingers from their hands to get their rings. A large number of responsible citizens were sworn in as guards, and they were instructed to shoot on sight any ghoul seen mutilating the dead. One of the guards told me that he shot twelve men in the act of robbing bodies, and that more than one thousand ghouls were shot before body robbing could be checked. These despicable criminals were men who resided in that part of the city which had not been destroyed. The dead, when identified, were given proper burial, but thousands could not be identified and the corpses were tied to iron rails, placed on barges, and hauled out to sea and buried in the briny deep.

Notwithstanding the fact that 6,000 residents of Galveston had been hurled to death in a few hours by that horrible monster, a storm tide pushed forward by hurricane winds, Galveston was not dead. Preparations to prevent such a calamity in the future were started immediately, and the rehabilitation of the city progressed rapidly. Construction along the Gulf shore of a sea-wall, higher than the storm tide of September 8, 1900, was carried to completion, and the city was raised to the elevation of that tide mark. Galveston is now as safe from damage from the storm tides of tropical cyclones as any other coastal city.

San Felippe de Austin Commandery, Knights Templar, No. 1, of Texas, which I had served as Recorder for many years, had elevated me from that position to Eminent Commander, an honor which I shall always prize highly. Membership in the Masonic organization gave me strength in the hour of trouble. All my earthly possessions had been destroyed by the storm tide, but my three little girls were left to me. The Masonic bodies of which I was a member gave me financial aid. The Grand Treasurer of the Royal Arch Chapter of Texas received a letter from a mason in Baltimore, Maryland, inclosing a nice sum with instructions to give the money to Dr. Cline if still alive. I hesitated to accept the money, because I thought there might be some mistake, for I could not place the man. The Grand Treasurer wrote the man and asked him why he requested that the money be given to Dr. Cline. The man in reply said that Dr. Cline had rendered him a great service in Tampico, Mexico, in 1898. He had been unable to make arrangements to reach some timber interests up the river from Tampico, and added: "I met Dr. Cline who secured a boat for me and the journey was a success." I had not done any thing more for this man than I would have done for any other man who needed assistance. I have always found pleasure in being helpful to others regardless of their creed or their station in life. My reward has been millions of friends.

The winds and the storm tide destroyed my home which had been built with borrowed money. The lots of ground alone remained, and they were almost worthless. I owed a large sum of money which I had borrowed to build the house. Friends advised me to go into bankruptcy and wipe the debt out. None of my ancestors had gone into bankruptcy; they had always paid up and I was determined to follow in their footsteps and pay the loan off as soon as I could. The money had been borrowed from a lawyer who represented individuals and not a loan company. About two years later a partner in the law firm came to see me. He said immediately on his arrival, "I have not come to ask you to pay that loan; the people who loaned you the money, will, on account of the great service you rendered the people of Galveston during the storm, cancel the claim against you if you will pay about $250.00, the cost of a friendly suit to clear the title of the lots to them". This magnanimous offer was accepted and the debt was honorably wiped out.

CHAPTER XVII.

CHANGE FROM MEDICINE TO CYCLONES

MEDICAL climatology, which I had selected as a subject for special investigation to be carried on in my recreation time, proved an interesting study. I soon found that very little research had been given to this matter, for there were no publications to be found bearing on this important subject. The influence of climate and weather changes on man, especially in the treatment of the different diseases to which mankind is heir, has not received the attention the importance of the subject merits. Very few members of the medical profession are equipped to carry forward researches of this nature.

Weather changes, in differing degrees, take place in nearly all parts of the world. The atmosphere is the principal element in man's environment, it is the breath of human life. Civilization had its inception where the climate may be said to be stable. This was in Egypt where the weather changes less than in any other part of the world. Here there are no clouds and no rains, there is no temperature change of importance—just enough to form a deposit of dew. The River Nile furnishes the water supply. Its floods overspread the bottoms, and as they recede the husbandman sows the seed. He then has nothing else to do until time to harvest and gather in his grain. In ancient Egypt conditions were such that the farmers took no thought of clothing, housing, and such. Civilization spread to other regions where weather changes are an important factor. As a result the greater effort of man's life was to adjust himself to the atmosphere in respect to weather changes.

Legends relating to weather and weather changes which had their origin more than 6,000 years ago have come down to us in the form of proverbs without much change. The slow progress made in the study of weather is surprising. The barometer was not invented until 1643, A. D., and the special study of weather and its changes did not receive much attention until two hundred years later. (See Appendix A).

Nearly fifty years ago I read a paper before the Texas State Medical Association entitled, "The Climatic Causation of Disease", which was published in their proceedings for 1896. The following comments on the paper show the interest at that time in such studies. Dr. J. W. McLaughlin said—"The paper is certainly very interesting, and the somewhat definite relations it shows to exist between climatic influences and the prevalence of certain types of diseases is very instructive." Dr. H. A. West said—"I feel like offering a vote of thanks to Dr. Cline for this exceedingly interesting paper; especially as it is upon a subject which has received comparatively so little attention." Other comments were all in the same vein. I had been studying this subject for fifteen years and at that time contemplated making it my life work.

"Man proposes but God disposes." The Galveston hurricane changed my objective. The destruction wrought by the storm tide caused by the winds of the cyclone, and the appalling loss of life in other tropical cyclones from the same cause convinced me that with proper knowledge as to the cause of these storm tides the tremendous loss of life and property could in a great measure be prevented. I was certain that people, living in localities exposed to tropical cyclones, could be correctly warned and moved out of the danger zone before the storm tide reached them. The theoretical circulation of the winds in cyclones, according to accepted theories at that time, would not account for the cause of the death dealing storm tides that accompany tropical cyclones. I decided that I could be of greater service to humanity by determining what are the physical forces in the cyclones that develop the storm tides and by devising rules for use in forecasting and warning the public in advance of their arrival, than I could by continuing my medical investigations. I gave up my medical work and started the study of cyclones and storm tides.

CHAPTER XVIII.

TRANSFER TO NEW ORLEANS.

UNDER the progressive administrations of General Greely and Chief Willis L. Moore the weather service became very popular. The probabilities of what the weather might be, as issued for the public at first, soon developed into specific forecasts stating what the weather, temperature, and winds would be during the coming 24 to 36 hours. This service was soon recognized to be of value by shipping, agriculture, and in fact every kind of business. These interests all clamored for improved service, and every step forward resulted in further demands for improvement of the forecasts. Increasing public demand for more specific forecasts of weather conditions taxed the Central Office forecasters at Washington beyond their capacity to meet these demands. Chief Moore soon realized that forecast districts located outside of Washington would enable the Bureau to give the public an improved forecast service. The first of the new forecast districts was located on the Pacific Coast, with the main office at San Francisco. Another forecast center was located on the Great Lakes with the main office at Chicago.

The Galveston disaster of September 8, 1900, and the failure of the forecasters in the Central Office at Washington to give warning of the approach of the tropical cyclone impressed Chief Moore with the importance of establishing a forecast center for the Gulf States. He informed me that in recognition of my work in warning the people of the danger threatened by the approaching tropical cyclone of September 8, 1900, he had selected me for forecaster in charge of the Gulf District. The district at first comprised Texas and Oklahoma, but was later extended to include Louisiana, Arkansas, Mississippi, Alabama, and Northwest Florida. In addition to the duties of forecaster in charge of the new forecast district I was Section Director of the Texas Section of the United States

Weather Service, and executive administrator of the central station of the new forecast district.

Forecasts for Louisiana and New Orleans were made by the forecasters in the Central Office at Washington and telegraphed to New Orleans for distribution. These forecasts were so unsatisfactory that the leading newspaper of New Orleans, The Times-Democrat, ran a "deadly parallel" column. One side gave the weather forecast and the other the actual weather performance of that day. This showed the forecasts to be so inaccurate that it made the forecasts issued from Washington look like guess work. Chief Moore wrote me early in 1901, and told me that the situation in New Orleans was giving him much concern. He had sent an inspector to New Orleans who tried to persuade the editor to discontinue publishing the parallel columns comparing the forecasts with the weather. The editor told the inspector that more accurate forecasts would remedy the trouble. Chief Moore informed me that the forecast center would be transferred from Galveston to New Orleans and that other States would be added to the district. He said he would expect me to make forecasts satisfactory to the public so that the "parallel column" would be discontinued. He cautioned me to be prepared to move on short notice without letting the information about the change reach the public. He knew that the powerful Congressional delegation from Texas, if they learned of the proposed change, would endeavor to prevent the transfer of the forecast center from Galveston to New Orleans.

Early in August, 1901, I, with my three little girls, proceeded to New Orleans, reaching there before there was any publicity about my move. I assumed my new duties at once, and started making the forecasts for the new district. I was certain that I could make forecasts which would meet the requirements of the public and give them satisfaction. Col. Page Baker, the able editor and publisher of The Times-Democrat, was consulted soon after my arrival in New Orleans, and we discussed the matter of forecasts. He was told that I would make the forecasts for the New Orleans district in the future; that I wanted him to criticize the forecasts when they failed to meet the public requirements and to commend them when he thought they merited commendation. The parallel column of forecasts and weather was discontinued. During the thirty-five years I was chief forecaster at New Orleans my work was never criticized adversely in any of the newspapers, but many commendations were printed emphasizing the value of the forecasts of weather, storms and floods.

Sugar and trucking interests as well as cotton interests were contacted, and were asked to advise me of their needs in the way of forecasts and other weather information. Orange and truck growers and sugar planters used the forecasts of minimum temperatures successfully in protecting their crops.

The sugar planters of Louisiana were outstanding personalities as they had always been since the early settlement of the State. They were ardent supporters of the carnival and the French Opera, and took an active interest in all public sports. They probably continued to have their social poker games as related in one of Lorenzo Dow's stories of 130 years ago. The story showed how a thief in an effort to cover up his guilt con- |victed himself. There are a few men who can not play an honest social game. In the early part of the last century planters, after a bountiful harvest, would sometimes get together and play a poker game for large stakes. Such a game was going on in a room in a hotel in Baton Rouge when a sudden puff of wind blew all the candles out leaving the room in total darkness. When the candles were relighted all the money had disappeared. Every member of the crowd was a gentleman— it was a gentleman's game. No one dared accuse another for that would have resulted in a challenge to fight a duel, and duels were popular in those days. They sent the porter for the proprietor of the hotel. No one had left the room, and the hotel manager told them that no one would be allowed to leave until he returned with the sheriff. As he passed the hotel registration desk he noticed a man signing, "Lorenzo Dow". He exclaimed—"Are you the man who can read men's minds and tell what they have been doing?" Rev. Dow replied, "They say that I can." The proprietor then said you are the man of all men I want to see just now, and he related what had happened at the poker game. Rev. Dow asked the proprietor if he had a live rooster. Upon receiving an affirmative reply, he then told the proprietor to direct a servant to bring the rooster and another servant to bring the blackest of the large pots from the back yard to the room where the poker game had been in progress. The proprietor of the hotel introduced the Rev. Dow as the man who could point out the guilty one. Rev. Dow had the soot covered pot put down in front of the door where the men were awaiting the verdict. He put the rooster under the pot, looked at the men and said: "Gentlemen I want you to pass out by this pot one at a time. As you go by rub your fingers on the pot and when the guilty one touches the pot the rooster will crow!" He then closed the door and took the proprietor into the room just across the hall, instructing him as follows: "Examine the fingers of the men as they pass by you and when a man comes in with no pot black on his fingers call me." A few of the men passed out with black on their fingers. Then came one who made the motion of rubbing the pot, but when the proprietor examined him there was no black on his fingers. The proprietor called Rev. Dow who took the man by the hand and said: "Look, there is no black on your fingers, if you had touched that pot the rooster would have crowed—fork over that money." The man handed over the money.

CHAPTER XIX.

FLOODS IN THE MISSISSIPPI RIVER.

OCCASIONAL floods have always been matters of great concern to the inhabitants of the alluvial bottoms of the Mississippi River. The run-off from precipitation over an area of 1,240,050 square miles has to be carried through New Orleans and discharged into the Gulf of Mexico. Some 200 years ago the early settlers threw up artificial embankments, called "levees", as a protection against flood waters. At that time a levee about three feet in height was constructed in front of New Orleans. This was the beginning of the fight against Mississippi River floods which has been continued to the present day with only partial success.

Prior to 1828 we find historical records of floods. The following brief references to some of the more important floods are of interest:

1717—The first settlers encountered a flood.

1735—A flood of great proportions occurred this year. New Orleans was inundated.

1775—New Orleans was flooded by crevasse waters. (A crevasse is the name given a break in the levee through which the water flows and spreads over the surrounding country.)

1782—The Mississippi rose to a greater height than at any time in history. In the Attakapas and Opelousas country the inundation was general. The few high spots which the water did not reach were infested with deer and other wild animals.

1791—A flood caused crevasses which inundated New Orleans.

1796—The flood this year overflowed the banks of the Teche for some 60 miles above New Iberia.

1799—New Orleans was flooded by crevasse waters.

1811—Floods this year did much damage.

1813—Serious floods occurred this year.

1815—Much damage was caused by floods this year.

1816—The flood caused crevasses above New Orleans through which the water flowed into the city.

1823 and 1824—Damaging floods occurred in both these years.

Following the acquisition of the Louisiana Territory by the United States the levee system was extended rapidly. An engineering work was commenced which had for its object the protection from floods of the alluvial lands for several hundred miles along the Mississippi River. No engineering problem is so great but that the United States would undertake to solve it. The magnitude of the flood control project calls to mind a story worth repeating here. A citizen of the United States, who was something of a wag and made a living by his wits, was touring Europe. He was entertained by prominent people everywhere. But when his host would point out something of unusual interest the American would always "go one better" by boasting of a similar building or scenic spot in the United States that would far surpass any shown him. There were historical buildings in the United States that even surpassed the antiquities of Rome. A group of Italian noblemen who entertained our American planned to get even with him. They doped his wine with alcohol and made him dead drunk. They took him to the Catacombs and deposited him in a sepulchre with the bones of the dead all around him. Then they waited for him to sober up and to hear what he might say. In about three hours he crawled out of the tomb, stretched himself and said: "By thunder, it is resurrection day and the United States is up first!"

Gauges for checking the height of the water in the Mississippi River were erected at several places in 1828. From this time on we have a record of the comparative stages of flood heights. Some of the more extensive floods were as follows:

1828—Levees offered very little resistance to this flood which covered the alluvial bottoms generally. The stage on the gauge now in use at New Orleans was 15.2 feet.

1844—Crevasses in the right bank of the river flooded the Tensas Basin. The stage at New Orleans was 15.2 feet.

1849—Several crevasses occurred in the right bank of the river between Red River Landing and Donaldsonville. A crevasse in the left bank occurred at Sauve's Plantation, 17 miles above New Orleans and flooded the City. The stage at New Orleans was 15.2.

1850—Several crevasses occurred during this flood and considerable areas were inundated. A crevasse at Bonnet Carre on the left bank flowed for six months. The stage at New Orleans was 13.8 feet.

1858—This was considered the greatest flood of record up to that time. There were several crevasses and large areas were flooded. This flood was of great historical importance because for many years it was used as a basis for comparison of both previous and later floods. The stage at New Orleans was 15.1 feet.

1859—A large number of crevasses occurred in this flood and inundated a large area of the alluvial bottoms. The stage at New Orleans was 15.6 feet.

Flood crests at New Orleans during the 43 years from 1828 to 1871 never exceeded 16 feet, and the levees generally washed out with stages slightly above 15 feet. The flood stage on the Carrollton gauge (New Orleans) is 17 feet, which is 1.1 to 2.0 feet above the crest of the floods which occurred during the above period. Prior to 1874 the embankments or levees were not high enough or strong enough to carry the flood waters through New Orleans and much land was overflowed when the river reached the flood stage.

A movement to improve the protection of the alluvial lands along the Mississippi River from floods was commenced in

1874. The forty-third Congress, First Session, passed an act, the title of which reads: "An Act to provide for the appointment of a commission of Engineers to investigate and report a permanent plan for the reclamation of the Alluvial Basin of the Mississippi River subject to inundation." The Mississippi River Commission was created in 1879, and the building of the new levees was well under way in 1881. However, the levees were not strengthened sufficiently to meet the increased flood height until 1890.

- 1874—A flood this year caused numerous crevasses from Helena, Arkansas, southward. The great Bonnet Carre Crevasse occurred April 11, 1874; it developed to a width of 1370 feet and to a depth of 52.7 feet. This crevasse remained open for eight years, not being closed until 1882. The stage at New Orleans was 15.7 feet.
- 1882—There were a total of 284 crevasses this year and the entire basin was flooded. The stage at New Orleans was 15.0 feet.
- 1884—There were 204 crevasses in this flood, mostly small ones. A large crevasse occurred at Davis, 60 miles above New Orleans, which had a discharge of 140,000 cubic feet of water per second. The stage at New Orleans was 15.6 feet.
- 1890—Several crevasses occurred in this flood and much of the Alluvial Basin was flooded. The levees had not been strengthened sufficiently to carry the flood above 16.1 feet on the gauge at New Orleans.
- 1892—Several crevasses occurred in this flood but the levees had been strengthened until they carried a stage of 17.4 feet at New Orleans.
- 1893—This flood reached 17.4 feet at New Orleans with a much smaller volume of water than that of 1892. There was a crevasse in the left bank 90 miles above New Orleans.
- 1897—The levees below Red River Landing had been strengthened so that only one small crevasse oc-

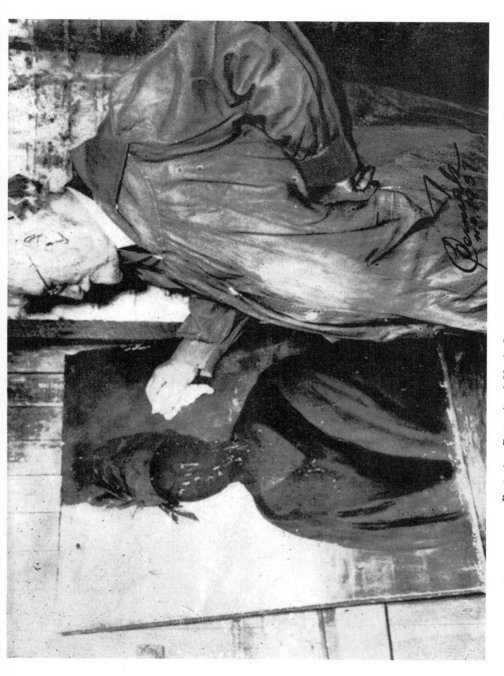

Restoring Portrait of Marie Laveaux

Photo by J. Cermak

curred between that place and New Orleans. The water was held within the channel between the levees and gave a higher gauge reading with a smaller volume of water than in previous floods when the levees gave way and the water spread out over the bottoms. The stage at New Orleans was 19.2 feet.

Prior to 1890 the levees were not strong enough to carry a volume of water that gave a stage above 16 feet at New Orleans. From that time onward the levees were raised and strengthened rapidly, and when the water was held between the levees in the channel of the river the stages were higher for smaller volumes of water. However, when a crevasse did occur the water had a higher head, the spread of the flood was more rapid and destructive than previously and the damage was often enormous. The stages reached on the several gauges depended upon levees carrying the water in the channel between the levees.

Warnings of the approach of floods from excessive precipitation in the Valley of the Father of Waters had assumed greater importance as the strengthened levees carried the water to higher stages before a crevasse occurred. No flood of importance had been carried through without crevasses. Forecasting these floods promised to furnish me with an opportunity to perform valuable service for the peopple of this section. Up to the time of my arrival the flood forecasts did not give definite information and consequently could not be used to any material advantage.

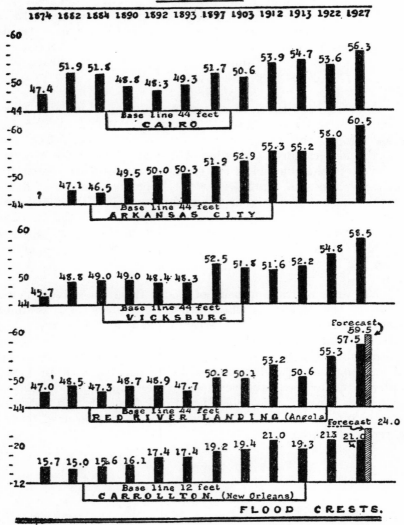

Increase in flood heights during 53 years as levees were raised and strengthened.

CHAPTER XX.

THE MISSISSIPPI RIVER FLOOD OF 1903.

STUDIES of floods in the Mississippi River which had been carried on from the time of my arrival in New Orleans and the conclusions reached were soon to be tested out in actual work. The Mississippi River started rising February 7, 1903, when the stage at New Orleans was 9.2 feet. The last previous important flood was that of 1897. The advices given the public in connection with that flood consisted of generalities. They were not specific forecasts, as can be judged by the following advice taken from the United States Weather Bureau Report on that flood:

> "March 15, 1897—The impending flood will prove very destructive in Arkansas and Northern Louisiana."

Advices of this kind could have been applied to any of the floods which had occurred in the history of the Mississippi River; for all of the floods had overflowed the low land along the River and seriously damaged crops. Here is another sample bulletin:

> "March 19, 1897—The floods in the lower Mississippi during the next ten days or two weeks, will in many places equal or exceed in magnitude and destructiveness those of any previous years and additional warning is given to residents of threatened districts in Arkansas, Louisiana, and Mississippi to remove from the region of danger."

Stages which would occur were not given in this bulletin. It does not state what specific areas were threatened with inundation. The topography of the terrain determines the areas from which people must move in order to escape flood waters. Forecasts of actual stages with dates on which they will occur are absolutely necessary to enable people to move out in advance of the floods.

The levees carried the flood in the river channel through New Orleans where a stage of 19.5 feet was recorded on May

13, 1897. The fact that the flood was carried in the river channel demonstrates that the volume of water was not as great as in many previous floods. There was no flood of consequence in the Mississippi River during the following six years. The fact that the flood of 1897 was carried through New Orleans without a break in the levees below Red River Landing made the engineers complacent and resulted in a failure to strengthen the levees. The wharves at New Orleans had settled to a point at which they would be overflowed when the river reached a stage of 17 feet on the gauge. With this situation in mind I studied carefully the flood outlook for the Spring of 1903.

Heavy rains over the Mississippi River Basin in February, 1903, continuing into March, flooded all tributaries and then the mighty Mississippi itself. This tremendous volume of water must pass by New Orleans, and a careful check was commenced on February 7th when the stage at New Orleans was 9.2 feet. Calculations showed that the volume of water in sight would reach 21 feet on the gauge at New Orleans if the levees could be made strong enough to keep the water in the river channel. The levees along the river front of New Orleans would have to be raised from three to five feet to carry a 21-foot flood. Unless raised and strengthened to hold the river in its banks the flood waters would spill over the top of the levee and rushing down a slope of 20 feet in two miles would cause disaster. Such an overflow, if allowed to occur, would washout the streets, undermine buildings, and cause damage that could not be estimated in dollars. Three to four weeks strenuous work would be required to raise the levees to a sufficient height to prevent such an overflow.

Long range flood warnings stating specific stages and the time they would occur in the Mississippi River were issued and distributed to the public on March 9, 1903; one bulletin read as follows:

"The Mississippi River is now within 1.4 feet of the highest water ever recorded, and the Atchafalaya within 1.0 foot of the highest water on record. These rivers will continue to rise slowly during the next three or four weeks, and all interests are advised to prepare for a stage of 21

THE MISSISSIPPI RIVER FLOOD OF 1903

feet at New Orleans and a corresponding rise in the Atchafalaya, if levees hold volume of water now coming."

(Signed) Isaac Monroe Cline.

This warning was sent by telegraph or mail to every town and village in the Mississippi Valley from Vicksburg, Mississippi to the Gulf of Mexico. All warnings issued had to be telegraphed at once to the Chief of the Weather Bureau, Washington, D. C. This was making history in flood forecasting, but it did not meet the approval of the officials at Washington. An order came back immediately by telegraph, saying:—

"Deemed advisable for you to withdraw your warning for 21 feet until certain that stage will be reached."

(Signed) Frankenfield.

This was answered as follows:

"If levees hold volume of water in sight that stage is certain to occur. Is it not better to warn the people of approaching disaster sufficiently in advance to enable them to prepare to meet it than to let them be caught unprepared."

(Signed) Cline.

Appproval of the warning for 21 feet was not received, my last telegram being ignored. This implied that the order to withdraw the warning for 21 feet must be obeyed.

The people living along the lower Mississippi would suffer enormous damage unless they had warning in time to enable them to prepare against a stage of 21 feet. I was serving the public, so I disobeyed the orders from the Chief of the Weather Bureau by repeating the warning of an expected 21 feet at New Orleans within three to four weeks every morning in the regular river bulletin. None of my important warnings had ever failed, and I had issued some exceptional ones. I did not let the public know that I had been ordered to withdraw the forecast for 21 feet at New Orleans, because that would have made the public doubt my forecast and prevented action to guard against the coming flood.

Many persons in New Orleans, including some members of the Levee Board, expressed the opinion that a stage of 21 feet would not be reached by this flood. However, the Levee Board commenced strengthening the levees.

Such slow progress was being made in preparing for the stage of 21 feet that the editor of the Times-Democrat consulted me concerning the outlook and on March 15th published the following editorial:

> "It must now be clear, however, that the Board (Levee) should not relax its efforts to strengthen the levees in this district. The river which is already high, will certainly rise even higher within the next fortnight. Indications are now plentiful that the waters, which now touch the gauge at 19.3 feet will, before April 1, go as high as 21 feet. In its official report the Weather Bureau has warned the public to prepare for such a rise. This warning can not be disregarded with impunity."

This stirred the citizens to action and several thousand workers were put to raising and strengthening the levees. There still were some who did not think a 21-foot stage was even probable. The editor of the Louisiana Planter, who on March 14, published an editorial stating that the river could not reach 21 feet at New Orleans, published another editorial on March 21, as follows:

> "We have therefore not yet reached the flood height of 1897 at New Orleans, and as we stated in our issue of last week we are quite skeptical about the river here reaching the 21 foot high water mark so confidently urged by the Weather Bureau, and further than this we are led by our observation to doubt the making of a 20-foot record at New Orleans this season. Still this latter may come."

This editorial did not influence the Levee Board. They hurried the work of strengthening the levees; and all levees, which were not high enough to stand up under 21 feet of water, were raised and reinforced with sacks of earth and planks. The embankments along the greater portion of the river front of New Orleans were raised from 2 to 4 feet, and in some places 5 feet, with temporary levees. On March 15, there was a crevasse in the levee of the left bank, fifty miles below New Orleans, which acted as a spillway and lowered the final stage at New Orleans. On March 26, a large break in the levee occurred in the right bank, forty miles above New Orleans, overflowing the Hymelia Plantation. It was known as the Hymelia crevasse. One thousand laborers were put to work immediately in an effort to build a crib dam across the 3,000-foot opening through which the water was flowing with a depth of more than 20 feet. The discharge of water through the crevasse was retarded but not stopped by the cribbing, and the outflow of water reduced the final stage at New Orleans. When the crevasse occurred at Hymelia, March 26, seventeen days after the warning for 21 feet at New Orleans was issued, a stage of 20.7 feet was recorded that afternoon. If the levees had held and carried the water by New Orleans the stage would have exceeded 21 feet. Notwithstanding the great volume of water rushing out through the Hymelia crevasse, which acted as a spillway, the flood water in front of New Orleans stood two to four feet against the temporary sandbag levee for nearly three weeks. The sandbag levee constructed as a result of the warning for a flood of 21 feet was all that prevented for nearly four weeks, the water from flowing down through the business section of New Orleans.

Levees which held the water out of New Orleans for nearly four weeks would not have been constructed if I had obeyed the orders from the Chief of the Weather Bureau and withdrawn the forecast for a stage of 21 feet at New Orleans. The water would have surged, with a head of 2 to 3 feet, down a slope of 20 feet in two miles. Such a flow of water would have washed out streets and undermined buildings to an extent that is impossible to conceive the enormous damage which might have resulted. The accurate warning, promulgated daily against orders, for more than three weeks resulted in precautions being taken which enabled the people of New Orleans to proceed with business in an ordinary way. The public had been given a service the value of which could not be estimated, but I soon learned that punishment for disobedience of orders would be meted out to me.

Sandbag levee heights can be judged from an illustration shown here.*

It shows the water over Cromwell Wharf at the foot of Toulouse Street. There were several miles of sandbag levee like that shown in the illustration. (See Plate II.)

---

* From United States Weather Bureau Bulletin, M, THE FLOODS OF THE SPRING OF 1903 IN THE MISSISSIPPI WATER SHED.

CHAPTER XXI.

ORDERED TO HONOLULU BUT ORDERS REVOKED

NINE months had elapsed since I disobeyed the order to withdraw the warning for a flood of 21 feet in the Mississippi River at New Orleans. The warning was justified and proved of inestimable value to the State of Louisiana. However, the fact that a great service had been rendered did not count for anything when in rendering that service a Bureau order had been disobeyed.

In February, 1904, I received an official communication saying that my forecasts for Arkansas and Oklahoma were not up to requirements, and that I would soon be transferred to Honolulu. Mr. James Berry, Chief of the Climatological Division of the United States Weather Bureau, arrived in New Orleans direct from Washington just at this time. When I showed him the letter he was much surprised and said he had not heard a whisper of it in Washington. He pronounced the proposed transfer an outrage and said he would fight it as soon as he returned to Washington. I did not let the public know anything about the proposed transfer. In fact, both the order to withdraw the warning for 21 feet of flood water and the order to go to Honolulu were never given to the press, and I am now releasing the facts about them for the first time. In previous instances, important forecasts and warnings were issued on my own initiative without authority when it was evident the public needed warnings which the forecasters in the Central Office at Washington had failed to issue. These failures I felt should be laid before the Chief of the Weather Bureau in Washington so that he could see the weaknesses in the service in order to correct them. I could do this without making invidious comparisons, for the Central Office had failed to issue important warnings when they meant much to the public. Indictments against the forecast service out of Washington, covering twenty-five letter pages, were sent to

Chief Moore. The first of these indictments was for the Bureau's failure to issue hurricane and emergency warnings against the destructive hurricane which almost destroyed Galveston—I had violated existing orders, stepped into the breach and warned the people of approaching danger. The Chief was told, for the first time, that I had diplomatically prevented an investigation of the failure of the forecasters in Washington by steering the President of the Associated Press away from such action. (See chapter on Galveston hurricane). His attention was called to the fact that I had been sent to New Orleans because of severe and just criticism by the New Orleans Press condemning the poor forecast service from Washington. Along with many other things his attention was called to an editorial which appeared in the New Orleans Times-Democrat, April 12, 1903, which not only commended my flood warnings but also my warnings to sugar growers and other agricultural and business interests. Extracts from this editorial are of interest here:

WEATHER BUREAU AND HIGH WATER

The value of the Weather Bureau to the country is being better recognized with every year. We of Louisiana are under special obligations to it. No crop is in greater danger from a sudden freeze than sugar cane and a few days' warning of an approaching freeze frequently means the saving of millions of dollars on the sugar crop. * * * By this splendid system adopted for the dissemination of cold wave signals, every planter in Louisiana is informed within a few hours of the prospective freeze and is thus given ample time for preparation and he is able to windrow his cane or otherwise protect himself.

* * * * *

We have been placed this year under another obligation to the Weather Bureau for its high water news and predictions. It has kept the people of the lower Mississippi well informed of what they may expect in the way of high water, and its predictions have subsequently been verified by the facts.

The day the high water would reach New Orleans was stated with remarkable accuracy, for it was between three and four weeks after the warning that the wave crest reached here.

ORDERED TO HONOLULU BUT ORDERS REVOKED

That the warnings had a good effect, like that of an approaching freeze, no one can doubt. It let the levee boards, planters, and public generally know what to expect in the way of high water, and warned them to prepare accordingly and they did prepare, raising levees to the height sufficient to withstand the flood which the Weather Bureau warned was coming. In this way it contributed not a little to the energetic and generally successful campaign against the flood carried on this year.

Many of the higher officials of the Weather Bureau in Washington recognized that I had rendered distinguished service, and took up the fight in my behalf. However, those who were trying to get me out of the United States made strong efforts to get me transferred to Honolulu. My friends told Chief Moore that if he transferred me from New Orleans and another great flood in the Mississippi River should occur without warnings equal to those I had issued for the flood in 1903, he would have plenty of trouble on his hands. The failure to issue flood warnings would be charged to him. They reminded him that I had been sent to New Orleans because of complaints of poor forecast service from Washington for that section, and that I had made forecasts which satisfied the public in every particular. Those seeking my transfer tried to convince Chief Moore that my flood warnings of 1903, were accidental and that I could not make such forecasts again. My friends replied, "Then you had better keep Cline in New Orleans, and let him make the mistake before you transfer him". This made Chief Moore consider the matter from a different point of view. My friends won out, and the instructions for my transfer were revoked. Chief Moore in his letter to me withdrawing the transfer order said:

> I was told by a subordinate that your forecasts for Arkansas and Oklahoma were not up to requirements and believing him to be honest I took the action ordering you transferred. The intercession of your friends caused me to call for the daily weather maps and I made forecasts for the months in which your forecasts were criticized and the official verifier make verifications. My forecasts were no better than yours and I consider myself a good forecaster.

Chief Moore cautioned me to request the newspaper men to cease complimenting my forecast work so highly, as though I were able to control the press! The forecasters in the Central Office at Washington recognized that my work was superior to theirs, and they wanted to get me out of the United States so that their work could not be compared with mine to their disadvantage.

Three months later I received a letter signed by H. C. Frankenfield, Chief of the River and Flood Division of the Weather Bureau, stating that the Chief of the Bureau directs that in the future forecasts for flood crests were not to be issued for periods longer than ten days in advance of the date the crest is expected to occur. I filed this letter without acknowledgment; for in the event another flood of importance should be indicated I intended to see that the public receive timely warnings. I would prepare a warning along the same lines as that of 1903, and telegraph it to the Chief of the Bureau in Washington, followed by a recommendation that such a warning be issued and distributed to the public. The officials in Washington would either have to issue the warning I had prepared or authorize me to issue the same. This course was followed in later years as will be seen in another chapter.

Something very unusual happened a few years after this flood. The official in charge of an important Weather Bureau station disappeared and no trace of him could be found. A few days after his disappearance the body of a man was found in a river near by, identified as that of the missing weather official, and buried. About two years later imagine my surprise when the man, who had been pronounced dead and buried, walked into the Weather Bureau office at New Orleans and asked permission to send a telegram to the Chief of the Weather Bureau to ask for reinstatement in the service. In less than half an hour the man's wife appeared in the office and showed no surprise. Nothing was ever learned concerning the man during his two years of absence.

CHAPTER XXII.

ART COLLECTOR.

SALARIES of Weather Bureau officials around the year 1900 were not large. The Galveston storm had left me at the age of 40 with my finances impaired. However, my credit was rated high. Recreation time had money value on which I determined to capitalize. I cast about for something I could do in my spare time that would be remunerative while not interfering with the discharge of my official duties. In my early life I had developed an interest in the study of art. Art now appeared to be a promising field, and one that would bring me in contact with cultured and refined people. The utilization of my recreation time to collect art objects would broaden my personal acquaintance with the people whom the weather service was created to serve. This acquaintance would enable me to serve the public more effectively and successfully.

Colonial Louisiana was populated mostly by French and Spanish people, who with artistic temperament, combined the pursuits of industry with the pleasures of collecting art. The wealthier classes brought with them to this country their art treasures, fine paintings, porcelains, enamels, bronzes and furnishings. The purchase of the Louisiana Territory by the United States was the signal for a large migration to the new territory from the eastern and northern States. Among these were some of the most cultured and wealthiest citizens of the country, who looked upon Louisiana as a land of promise which offered great opportunities to those with capital and initiative. These people brought with them their family portraits, painted by the best artists of the time, and other art treasures and heirlooms. Thus this section was settled by people who patronized the fine arts.

The industries of this part of the South grew rapidly and within a few years Louisiana was the richest part of the United States. Culture kept pace with the rapid strides of industry and reached its zenith between 1840 and 1860. From about 1816 until the War Between the States New Orleans was the mecca for portrait painters from the eastern States, and many prominent European painters came here to paint portraits of the rich and cultured people of this section. Among the distinguished painters from the eastern States and Europe, I will mention, Eliab Metcalf, John Wesley Jarvis, John Vanderlyn, Henry Inman, Matthew Harris Jouett, William Jewett, J. G. L. Amans, G. P. A. Healy, Samuel L. Waldo, Thomas Sully, A. Rinck, Jean Joseph Vaudechamp, Henry Bird. There were others of equal ability. Alfred Boisseau, a noted French genre painter, has left us his painting, "Indians Marching", in a landscape on the Tchefuncta Creek, painted in 1847, and exhibited at the Paris Salon. The same year history tells us that Vanderlyn exhibited the panorama, "Versailles", at the plantation of Valcour Aime, in Saint James Parish. Jarvis spent his winters in New Orleans from 1818 until 1832 painting the portraits of prominent men and their families. It is said that he took in upwards of $6,000.00 every year he visited New Orleans. Jarvis entertained his patrons, and it is of record that just before his departure from New Orleans one Spring he invited twelve of his friends to dinner. One of the crowd proposed that each guest put a substantial sum of money in a bowl on the table. Then each man would tell the biggest lie he could conjure up, and the money would go to the person voted the biggest liar. Jarvis was toastmaster, and, commencing on his right, he asked each man to do his best at telling a lie. Jarvis' turn came last and he said, "Gentlemen, thank you. I believe every word that each of you have said; cast your ballots according to your judgment". They voted Jarvis the greatest liar.

The interest displayed in art by the people of this section attracted the attention of art collectors in Europe. Several Italian Noblemen conceived the idea of bringing a fine collection of paintings by the Old Masters to New Orleans for the purpose of establishing a National Art Gallery in this city. Accordingly they brought 338 fine examples of the works of the Old Masters, selected by experts from the finest collections in

Europe. The plan for the establishment of a National Art Gallery fell through, and the paintings were sold at auction in the Ball Room of the St. Louis Hotel in 1847—the first big art auction in America. I have the catalogue of this sale of the largest collection of paintings ever exhibited in the United States up to that time. While a National Gallery was not established in New Orleans, the paintings were dispersed throughout the United States, and formed the nucleus for art galleries in different parts of the country. The well-known American painter, Worthington Whitridge, in a letter written shortly before his death stated that he got his first inspiration and impulse to be a painter from some of these paintings which found their way to the museum in Cincinnati.

Another important collection of paintings and other art objects was that of James Robb. This collection was sold at auction on February 26, 1859. I have the catalogue of this sale which contained a number of paintings that Robb had bought from the collection of Joseph Bonaparte.

Choice paintings and other works of art appealed to me. I bought paintings, porcelains, bric-a-brac, and small pieces of furniture soon after my arrival in New Orleans as a hobby. My recreation time was devoted to this hobby. I used my annual leave in short periods of from two to six hours at a time so that I could attend auction sales and visit second hand and antique stores. Many of the paintings were damaged and had to be repaired before they could be displayed. Two fine old paintings which needed repairs were given to a man bearing fine recommendations, who had just arrived from New York. He cleaned them with acids, removed the glazes and then reglazed them. He destroyed the originality of the paintings, and this made them worth no more than copies. After that no restorer touched my paintings. I did my own restorations, and the originality of the paintings was preserved except where paint was missing. The most exacting collector could not find fault with my work.

Credit enabled me to buy on time payments, and I soon was in debt to all antique and second hand stores in the city. When I bought from private individuals the purchases were for cash. When people wanted to dispose of art works and the price was reasonable I paid it without question. Frequently I came in contact with persons who had suffered adversity and they would want me to put a price on their paintings. In such cases I would not put a price on the paintings, because after I had restored them they would be more valuable and the persons who had sold them to me would feel that I had not paid them enough. I would suggest to such people that they have Prof. Ellsworth Woodward, Director of Art, at Newcomb College, to appraise the paintings, and I would then decide whether or not I could use them. Where people were in needy circumstances Prof. Woodward would not make a charge for the appraisal. If his appraisal was satisfactory I would buy the paintings. Occasionally when people were in distress I would pay more than the paintings were actually worth to me in order to give them financial relief. Paintings which did not come up to my expectations, after being cleaned and restored, were sent to auction and sold for whatever they would bring. I always realized enough from the sales to pay me for cleaning and restoring. Paintings with merit were retained in my collection. I borrowed money from the Hibernia National Bank, and the bank was kept informed as to my general indebtedness.

Financial crises are dangerous to any one, especially to a man involved in debt. In April, 1907, Mr. Charles Palfrey, Cashier of the Hibernia National Bank, called me into his office. He informed me that a financial panic would be felt in New Orleans the coming winter. He told me that I need not worry about what I owed the bank but that I had better pay all my other indebtedness, otherwise I might be pressed for payment and suffer serious losses. Plans were made at once to sell the greater part of my art collection at auction. The sugar crop was promising and September was selected as the month for the sale. Early American portraits, Keith landscapes, and some choice porcelains and bric-a-brac were reserved as the nucleus for a future collection. There were three hundred lots to be sold. Jac H. Stern was employed to conduct the sale with the understanding that the sale must be completed in one day, for I did not want to be bothered with

Plate II. Showing emergency sandbag levee nearly 5 feet in height at the Cromwell wharf, foot of Toulouse Street (New Orleans). The flood waters stood at a depth of 3.5 feet, above the wharves, against this levee for nearly a month.

a two days' sale. The day selected for the sale proved ideal and the last lot was sold at 6:30 p. m. There was not a limit on any article. Some paintings for which I had paid four or five hundred dollars would go for 100 to 200 dollars. Other paintings which had cost one or two dollars, which I had relined and restored would bring one hundred to two hundred dollars. The sale was a great success. Debts were paid off and I had money to start buying another collection. I also had some good paintings, porcelains, and other art objects as a nucleus on which to build another collection.

TORNADO IN NEW ORLEANS.

Tornadoes with the funnel-shaped clouds are seldom seen in New Orleans except small ones which appear over the Lake in the form of water spouts. October 5, 1906, about 8:15 a. m. a well defined tornado was observed in the vicinity of Soniat and Bellecastel Streets. The funnel-shaped cloud, from which the disturbance extended down to the ground, traveled towards the northeast and disappeared in the vicinity of Clara Street and Tulane Avenue. The path of the destructive area was less than 300 feet in width at its widest point. Three persons were killed and it caused $300,000 damage. Notwithstanding destruction and death stalked there were amusing incidents. Our colored servant on her way to work was passing a graveyard which was in the path of the tornado. She heard a roaring noise and looking over the graveyard she saw the tornado pitching the tombstones about like feathers. She came running at full speed into the house with fright, screaming "Lawdy de judgment day am here for as I passed the graveyard I hear a great noise and look 'round and see the dead risin' and throwing the tombstones up in the air like straw."

CHAPTER XXIII.

CARE OF OIL PAINTINGS

MANY people own family portraits and other paintings by distinguished artists. I offer a word of advice on how to take care of paintings and protect them from injury. I have given this subject careful study during the forty years in which I have been collecting paintings.

Paintings are of a delicate nature, and if not properly cared for they deteriorate rapidly. Unless stretchers are frequently keyed, and the paintings occasionally cleaned by an expert, the canvas becomes loose and the paint cracks and scales off. Paintings should never be wiped with a damp cloth. The dampness will enter the small cracks, which appear in nearly all paintings, and cause the paint to become loose and drop off. The mediums used for the conveyance of the colors are affected by temperature changes. Paintings should never be hung over a radiator or a fireplace because too much heat causes the paint to crack and scale.

There is something of merit in nearly all paintings; in fact there is sufficient merit in any painting to justify its preservation. The majority of people know very little about the merit of paintings, but every one knows what pleases the senses. The masses are waking up to the fact that there is a charm in viewing paintings that pleases in a way which they can not explain. Paintings give people who have only a slight knowledge of art something they can not get in any other manner. Painting is the language in which the artist expresses his ideas. Works of artists which have come down to us through centuries give us information which could not have been communicated to us in any other manner, and their "language" should be preserved in its original form as nearly as possible.

Collectors of fine paintings are exacting about the condition of paintings and the manner in which they have been cared for. A noted English art collector, the author of Smith's

Catalogue Raisonne of Art (published in nine volumes, 1831 to 1839), was not complimentary to the restorers of paintings of his time. He stated that many restorers cleaned paintings so closely that glazes which the artist had used were taken off, and the paintings reglazed. This operation destroys the originality of the paintings and makes them different from the artist's work although still bearing the artist's name. He emphasized the fact that a thin coating of the old varnish should be left on the paintings even though some dirt specks remain visible, and that all dirt can not be removed without removing some of the original paint. This must be avoided. I have found that the same criticism of restorers that Smith made 100 years ago still holds. I have seen many fine paintings cleaned so closely that the original quality and the personality of the artist were completely obliterated. (See preceding chapter.)

The study of art in early life and my experience with restoring paintings made me resolve to restore my own paintings in the future, and to adhere to the cautions and suggestions made by Smith a long time ago. I have cleaned some two thousand paintings during the last forty years. Many of these have passed into the finest private art collections in America, and many are now in important museums. Many of the paintings which I handled had been in the hands of incompetent restorers. I found much repainting or overpainting which never should have been done. In many instances paintings had been cleaned with strong chemicals, and so much of the paint removed that they could not be restored to their original condition. In other instances the restorer had not removed the old varnish but had painted over it. In such cases the over painting was removed and the painting restored to its original condition. Many fine paintings by great masters have been ruined in cleaning.

Paintings which were sold years ago for several thousand dollars, have sold in recent years for less than one thousand dollars. The loss in value was due to injury resulting from the work of incompetent restorers in the meantime. Some of the greatest painters got their loveliest effects by glazing, and when this is removed, the originality of the painting is destroyed. Where the paint is gone from the canvas, the missing part should be filled with color to match the original color as nearly as can be done, and there should be no glazing around the repaired place to bring the colors together. The repaired part should be visible on close examination, because the value of a restored painting depends on the amount of original paint still on the canvas.

In restoring paintings it is often necessary to reline the old canvas on account of holes or its decayed condition which will not stand stretching. This work requires experience, great care, judgment, and discretion. The proper way to do this is to seal a new canvas to the back of the old canvas with a medium that will not blister or develop bubbles. I find a great many relined paintings full of blisters where glue and paste were used in tying the new canvas to the old one. Great care should be taken to avoid pressing out the brushwork and technique of the artist. This can easily be done by too much rubbing in fastening the new to the old canvas. Here also care should be taken to preserve the painting in its original condition, even to the preserving of all that remains of the old canvas.

In studying the methods of restoring paintings I have read where some restorers remove the old canvas thread by thread and put new canvas directly on the old paint. In my judgment this cannot be done without injuring the painting. The threads of the old canvas tie into the artist's technique; therefore, when the threads of the canvas are removed, the brushwork is broken down and the pristine beauty of the painting is destroyed to a great extent.

CHAPTER XXIV.

ORIENTAL BRONZE COLLECTION.

ORIENTAL Art always held a charm for me. The reasons for this I could not explain unless it is because this art possesses unusual beauty of line and form. One of the finest collections of small Oriental bronzes to be seen in America is in the room bearing the designation "CLINE COLLECTION" at the Isaac Delgado Musem of Art in New Orleans.

Accidental—yes, this was an accidental collection. I did not have to visit the Orient to assemble a collection of 137 choice antique Oriental art objects. In November, 1910, Harry Fitzpatrick, auctioneer, advertised a sale of unclaimed freight held for storage charges in the Southern Pacific storage depot on the river front opposite Jackson Square. Bronzes were listed among the articles to be sold. However, I assumed that in such a sale of unclaimed freight there would not be anything of interest to me and I did not attend the sale. The following day Fitzpatrick had an auction sale at a residence in the Garden District of New Orleans. One day's annual leave was taken to attend this sale. I made it a practice when I attended auction sales to go early so that I could examine articles in which I might be interested before the arrival of the crowd. On this occasion as I entered the house fourteen Oriental bronzes in the forms of chimerical animals and birds attracted my attention. I was confident they were not part of the belongings in that house, and I inquired of the bookkeeper where those bronzes came from. He told me that some ninety bronzes of that character, contained in a hogshead barrel, were sold in the unclaimed freight sale in individual lots and that Fitzpatrick in bidding against his customers had these left on his hands. He had brought them to this sale to work them off. The book-keeper was asked for a list of the people who bought the other Oriental bronzes; he gave me the information and the number of bronzes each had bought. I then told him to buy the fourteen that were on sale for me without putting a limit on them; I knew they would not bring their value. I located the addresses of the persons who had bought the other bronzes, from the City Directory, and started on the hunt. When the sun set that evening I was the owner of sixty-five of the Oriental bronzes, quite a fine collection for a starter in one day's work.

Interest in the bronzes became very marked when it was learned that I had purchased them because I was considered a critical buyer. The purchase became a matter of conversation throughout New Orleans. Soon after I had made the purchase, the Japanese Consul telephoned me that Prof. Fukushimi, a noted Oriental Scholar and Professor of Archeology in the University of Tokyo, was in New Orleans en route to London, to supervise a Japanese exhibit at an international exhibition. He asked permission to bring Prof. Fukushimi to see the Bronzes. Upon viewing them he expressed amazement, saying that the collection contained many of the rarest of the Chinese bronzes, some of which were more than three thousand years old. The hogshead of bronzes had been shipped from China during the Boxer uprising prepaid, and consigned to a mark. The party with the identification never appeared to claim them and they were sold ten years later for storage charges. Prof. Fukushimi was much interested, and told me he expected to come by New Orleans on his way back from London, and asked permission to examine the bronzes at that time. In May 1911, the Associated Press carried a news item from a town in Virginia announcing that Prof. Fukushimi had died at that place after undergoing an operation. So I did not get a further critical evaluation of the bronzes from him.

Interest in art had developed in New Orleans to an extent that Isaac Delgado gave the City a museum of Art. The museum opened in November 1911, and the bronzes were loaned to the museum for an indefinite period. Frederick Moore, representative of the Associated Press in China during the war between the Buddhists and Toasts (1912-1914), was on a visit to New Orleans. We met in Farish's Art Store, and my collection of Oriental bronzes was mentioned. Moore remarked that the Buddhists and Toasts were destroying each other's temples, and that fine old bronzes could be bought for the value of the metal. Moore and Farish were close friends, and I suggested to Farish that he give Moore an order for fifty temple bronzes, and that I would take them on arrival unseen and give Farish his regular profit on such goods. Moore sent Farish fifty fine Oriental bronzes, and I added these to the collection in the Isaac Delgado Museum of Art.

One small cloisonne vase of the Ming Dynasty in the collection represents the withering of a romance. A United States Army Officer, who was stationed in the Royal Palace in Pekin during the Boxer uprising, sent this vase from the Palace to a girl to whom he was engaged to be married. The girl sold the vase to me and married another man.

The Second Pan-American Scientific Congress was held at Washington on December 27, 1915. I was a member of that Congress, and during my attendance I found a collection of Oriental bronzes which a Mr. Osgood had bought in China during the unsettled conditions under the same circumstances Mr. Moore had obtained his. I bought thirty of the choicest bronzes of the Osgood collection. I gave a few to my children and added the others to the Collection in the Isaac Delgado Museum of Art, now owned by the Museum.

CHAPTER XXV.

ANOTHER MISSISSIPPI RIVER FLOOD.

THE heavy discharge of water into the Mississippi River from the extensive drainage area of the upper valley, because of the excessive rain and snow during the winter of 1911-1912, created a problem for the flood forecaster. An estimate of the amount of flood water that would enter the Mississippi River required careful calculations and a thorough study of the run off. The height and strength of the levees along the banks of the lower Mississippi River that would have to hold this flood in the river channel were thought by some to be inadequate. Nine years had passed without a flood of importance, and it was evident that I would have an opportunity to prove to the forecasters in Washington that the verification of the forecasts for the flood of 1903 was not an accident as they had claimed when they endeavored to have me transferred to Honolulu. I would show them in this flood that long range forecasts of flood stages in the lower Mississippi River had come to stay. After the flood of 1903 the Central Bureau forecasters endeavored to put a stop to the issuance of long range forecasts. I was instructed by the Chief of the Weather Bureau to limit the period covered by my flood forecasts to ten days in advance of the flood crest. In the case of the flood of 1903, a warning for the crest only ten days in advance of its occurrence would have been useless—the public could not have prepared safeguards in ten days to protect themselves against the dangers from such a flood. As forecaster in charge of this district I considered it my duty to see that the public got timely warnings of any dangerous river conditions threatening them.

I determined to see to it that timely warnings were issued notwithstanding the limitations placed on me. The forecasters in Washington would either issue the flood warnings I would recommend or they would authorize me to issue them. I was

ANOTHER MISSISSIPPI RIVER FLOOD

certain they would not dare turn down a flood warning which I recommended, since I had disobeyed their orders in 1903 and proved my judgment to have been correct.

Rains in the early months of 1912, accentuated by conditions which preceded them, made it certain that the lower Mississippi River would experience one of the greatest floods in the history of the Weather Bureau. The Mississippi River had never carried a great flood without breaking through the levees below Memphis. Often the greater part of the flood waters was transferred from the river channel to the low lands along the river. Such conditions augmented the flood dangers below Vicksburg. Flood waters in the channel of the Mississippi travel at the rate of about 40 miles a day. When there are breaks in the levees the flood is in a great measure transferred from the channel of the river to the bottom lands. Timber and vegetable growth retard the rate of travel over bottom lands until the movement of the flood crest is only about 13 miles a day. Thus overland floods travel only one third as fast as channel floods. This complicates and intensifies the flood situation, especially when crevasses occur well in advance of the flood crest in the river channel, which has nearly always been the case. The crest of the channel flood and that of the overland flood come together frequently at some point down stream and give a higher stage from the point of junction to the Gulf than would have occurred if the river channel had carried the flood through. In some instances the increased height of the flood amounted to two to three feet. The levees during nine years without a flood of importance needed both raising and strengthening. The wharves along the front of New Orleans had been rebuilt since 1903, and were high enough to carry a 22-foot stage on the Canal Street gauge. Therefore this flood did not threaten the city of New Orleans, but the low lands below Vicksburg were threatened with destructive floods.

Conditions on April 3, 1912 indicated that a greater flood was approaching than that of 1903. The following flood warning was prepared and telegraphed to the Chief of the

Weather Bureau, Washington, with a recommendation that such a warning be issued:

> With the water now in sight, without any further rains, the Mississippi River below Vicksburg and the Atchafalaya will rise until the early part of May, and if the levees hold the river will reach stages, about 52 feet at Natchez, 42 feet at Baton Rouge, 33.5 feet at Donaldsonville, and 21.5 feet at New Orleans. These stages are in some instances about 1.0 to 1.5 feet higher than in any previous flood.
>
> (Signed) Cline.

Telegraphic instructions authorized me to issue the flood warning as recommended. The flood waters came with tremendous volume. From Memphis to Vicksburg nine crevasses occurred in the right bank of the river, ranging in width from 500 feet to 3,400 feet, and one on the left bank 2,400 feet wide. The water rushed through these crevasses in the levees and formed an extensive overland flood which covered the bottom lands. Much of this overland flood traveled down over the low lands on the right side of the river until it reached the levees along Old River where the Mississippi River and the Atchafalaya are connected. Here the overland flood combined with the flood in the channel of the Mississippi and caused an exceptionally rapid rise in the water of that area. The levee along Old River could not be built up fast enough to keep the water from flowing over it, and 3500 feet of the levee washed out at Torras. The water from the overland flood and that of the river flood combined rushed through this crevasse like a bursted dam, and soon overflowed all the land between the Mississippi River and the Atchafalaya. This flood was augmented by another overland flood from the Hymelia Plantation crevasse 1,200 feet wide in the right bank of the Mississippi. The water was over the tracks of the Southern Pacific Railroad, and pilot engines went ahead of trains to make sure there were no washouts. I made a trip by train from Morgan City to New Orleans, and the water came nearly up to the fire boxes in the engines. People living in one story buildings had moved out. Those in two-story buildings moved every thing to the second story and remained in their homes. Many of those who remained in their homes owned a cow, horse, goat or chickens. Platforms built on logs formed rafts on which the animals were placed to keep them from drowning. These rafts were inclosed with fences and tied to trees; the owners used canoes to get around to the platforms and care for their live stock.

The Mississippi River as a result of the overflow from the Torras and Hymelia crevasses covered the entire land area between New Orleans and Morgan City. The river reached a stage at New Orleans of 21 feet May 9th, 21.4 feet May 12th and the wind backed the water up to 22 feet at 2 a. m. May 12th.

A complete report on the flood was prepared and sent to the Central Office of the Weather Bureau at Washington. The warning was mentioned in a bulletin published some months later but long range forecasting was not referred to.

Bulletin "Y", U. S. Weather Bureau, Washington, 1912, gives the total value of property saved by the warnings for "The Ohio and Mississippi Floods of 1912", at $16,180,000 of which $10,000,000, nearly two-thirds of the total savings, was in the area covered by the long range flood forecasts issued from New Orleans, where the people had four weeks to prepare for the flood. If I had submitted to the order from the Chief of the Weather Bureau and limited my forecasts to ten days before the occurrence of the crest of the flood the loss would have been enormous.

XXVI.

POLITICAL ASPIRATIONS OF CHIEF MOORE.

WILLIS L. MOORE had been Chief of the Weather Bureau for seventeen years (1895-1912), at the time of this incident. He was considerate and showed good judgment in all respects during the early years of his administration, but as time passed he came under influences that turned him against some of his best friends. When a station official performed work that attracted the attention of the public and was commended by the press, Moore frequently sent him to some part of the world where he could not render conspicuous service. Small-minded and envious men, who happened to be close to Chief Moore, were the cause for such action. They persuaded him to punish those whom they disliked because of their successful work in the interests of the public—work which these sycophants were not capable of doing so well. Moore was cautioned by such men to be on the alert lest some other official might get his place as chief. Loyalty to my bureau Chief under whom I was serving was a principle that I would never violate. There were some who tried to make Moore believe that I would use my successful forecast of floods and storms as a stepping stone to become Chief of the Weather Bureau, a position to which I never aspired. Those who suggested that such a thing might happen were men who had failed in forecasting important disasters. They hated to see me step into the breach without authority and succeed. Friends kept me posted on such matters, and this enabled me to act cautiously and be exceptionally courteous and diplomatic where these officials were involved.

New Orleans was a dumping ground for observers who were guilty of drunkenness and neglect of duty and whom it was necessary to discipline. These men were inefficient as assistants, and this made it more difficult to keep the work of the station up to a high standard. The assignment of such men would be accomplished by a letter of advice stating that my firm but gentle discipline might bring these men around, to make useful men of them. Sometimes the object was to improve the men, but as a rule the object was to give my station a bad record in dealing with personal problems. The men were either put on their feet or forced to leave the service. The observer was always given the benefit of the doubt so that the records would show that I had not dealt unfairly with him. The fact that these trouble makers were sent to my station so that I might be convicted of harshness in dealing with assistants, came out in later years when a new Chief of the Bureau was appointed, as will be shown in a later chapter.

Chief Moore became well and favorably known through the excellent forecast service given under his administration. He took credit for the successes attained which was proper, but he did not give due credit to those who performed the service. In later years of his administration he yielded to political intrigues. Political ambitions, resulting from bad advice given him by some of his favorite subordinates, led him into activities not considered in the best interests of the public service. His real friends advised him against his contemplated political activities and he turned against them. He aspired to the position of Secretary of Agriculture in the Cabinet in the event that Woodrow Wilson should be elected President in 1912. He and his weak advisers planned and organized a campaign, looking to this end, in which the leading officials of the Weather Bureau would be expected to take an active part, or else - - -

Atlanta had been selected as the meeting place of the American Association for the Advancement of Science, for the year 1912. Chief Moore decided to have a convention of Weather Bureau Officials at Atlanta at the time of the other convention. The Weather Bureau Convention was to discuss matters of importance to the weather service, but the real purpose proved to be the furtherance of Moore's political ambitions. The expenses of this convention were paid from U. S. Weather Bureau funds.

Early in September, 1912, I received instructions to attend the Weather Bureau convention at Atlanta, but there was nothing to indicate that politics was to be the liet motif of the convention. No information had reached me concerning Chief Moore's political ambitions. Most of my friends in Washington tried to discourage Moore from seeking the post of Secretary of Agriculture. Moore and his advisers thought that I would oppose the move and he kept the plans for the campaign from reaching me. A few days before I was to leave for Atlanta to attend the convention, Norman B. Conger, general inspector in the Weather Bureau, arrived in New Orleans, ostensibly for the purpose of inspecting my station. I told Conger that I was under instructions to attend the convention at Atlanta, and that I presumed that he would be through with the inspection by Monday, the day I was to leave. He replied, "You may not go to Atlanta." In the afternoon of the same day a Mr. Burns, Chief of the Printing Division in the Weather Bureau at Washington, came in, saying that he had come to inspect my printing plant. Two inspectors arriving at the station at the same time, and Conger's remark put me on my guard. I suspected that some sinister move had been planned to railroad me out of New Orleans on trumped up charges supported by two inspectors. Another notable flood warning, issued four weeks in advance of the flood crest for the unprecedented flood of April and May, 1912, had attracted much attention through the press and otherwise. I had no knowledge of Moore's political aspirations. I asked no questions, but kept my ears to the ground and my eyes open, prepared to meet any move the inspectors might make.

New Orleans was preparing to entertain the Farmers' National Congress which was to convene Friday, the day after the arrival of the inspectors. Conger informed me that it was my job to get the Farmers' Congress to indorse Moore for Secretary of Agriculture—rather a big assignment for a man who had not been informed previously that Moore was a candidate for that office. After thinking over the matter for a few minutes I said, "Mr. Conger, I am personally acquainted with the officers of the Congress. I will introduce you to Hon.

Harvey Jordan, and Dr. H. E. Stockbridge, and other prominent leaders in the organization. Since you know all of the details you can explain to them what Moore wants better than I can." Conger was flattered and pleased with my suggestion. Friday morning I introduced him to the officers and prominent members of the Congress and let him do the talking. When he returned to the office late in the afternoon he was very much pleased. Conger then showed me a telegram which he was sending to Moore, it said: "Cline can be depended on." The Farmers' Congress did not endorse Moore but Conger succeeded in preventing them from endorsing another for the place. Printing inspector Burns did not inspect anything. Conger accompanied me to Atlanta. Moore sent for me immediately on my arrival there. He gave me messages to deliver verbally to Jordan, Stockbridge and others asking them to support him for appointment to the position of Secretary of Agriculture.

On the last day of the convention Moore handed me a list of ten of the highest ranking officials of the Weather Bureau in attendance, saying to me, "Here is something I want you to do; call a meeting of the ten men, and arrange for a dinner for thirteen persons." The expense of the dinner was to be paid by the ten of us without any assistance from Moore. A fine spread was prepared and it cost us a good sum. While at the dinner Moore told me to request travel authority at Government expense to visit certain educational institutions in the South and talk in the interests of his plans. I did not promise but made it a point to keep so busy that I could not make the journeys he wanted me to make. I did not spend any Government funds in making journeys in Moore's interests. I was certain there would be trouble where Government funds were used for such purposes. Moore was pleased with the action of the Farmers' Congress and with what I did for him in Atlanta. Along with several other officials I was promoted soon after the Atlanta Convention.

CHAPTER XXVII.

THE EXPLOSION.

WOODROW WILSON was installed as President of the United States on March 4, 1913. Chief Moore was so certain that he would be appointed Secretary of Agriculture that he had selected Henry J. Cox of Chicago as Chief of the Weather Bureau, and Cox had come to Washington to take over. Moore got the shock of his life when President Wilson appointed David F. Houston Secretary of Agriculture.

Rumors of all kinds were going around by the grapevine. Moore then saw the handwriting on the wall but failed to interpret it correctly. Investigation of the use of the Weather Bureau personnel by Moore, for political purposes, was evident, but how thorough this might be no one could conjecture. Word went out confidentially from Moore's office suggesting the destruction of all correspondence concerning his effort to secure the appointment to the position of Secretary of Agriculture. It was thought that by so doing the evidence on which Moore might be removed could not be found. The suggestion that I destroy my correspondence on the subject reached me verbally but was disregarded. When Conger told me what was in the air I realized that they were playing with dynamite, but to have told him so would have meant my demotion and transfer to some distant station. Therefore I decided to exercise caution and play the game diplomatically. I wanted to retain my position. Every precaution was taken to safeguard myself against the explosion which I felt was sure to occur. All letters received from Moore or from any of his lieutenants bearing on the subject of his candidacy were kept in their original envelopes with copies of my replies and filed where no one else could get to them. These letters were my line of defense in case I should be charged with any wrong doing in this matter. The letters would prove definitely that I had not violated any laws and that I had not gone out of my way to advance Moore's cause except when specifically directed by him. Notwithstanding Moore had personally instructed me to request authority to make journeys at Government expense and speak in his behalf I did not request authority to make any journeys at Government expense.

On April 1, 1913, "All Fools Day", the dynamite exploded. Moore and some of the officials who had been his personal advisers were suspended from duty pending an investigation. Promptly at 9:00 a. m. that day an investigator from the Department of Justice walked into every Weather Bureau office where the official in charge was known to have been connected in any way with the candidacy of Moore for Secretary of Agriculture. The agent handed me a letter from the Secretary of Agriculture addressed to me personally. This letter directed me to aid the special agent in every way possible in the work on which he was engaged and to keep the matter secret from those over me and those under me. The investigator had a complete record of every move I had made in Atlanta, even to the hour I went into my room at night and the hour at which I came out in the morning. He had not learned of the Farmers' Congress incident at New Orleans. One of the first things the investigator did was to ask to see my correspondence with Moore. He was surprised when I handed him the complete file with letters still in their envelopes. He remarked that the other officials under investigation had destroyed their correspondence on the subject. We had evidently been watched all the time. He asked me why I kept the letters. I told him that I had kept them as a matter of self protection, as they were my only defense for use as proof that I had not been guilty of any wrong doing. He remarked that I had done Moore a real service in keeping these envelopes. He was charged with having used the frank in his correspondence in carrying on his campaign for the office of Secretary of Agritulture, and these envelopes all having postage stamps showed that he did not use the frank for this purpose. The stamped envelopes disproved that charge. I told the investigator that in my judgment, Moore had the best interests of the public service at heart, and that he was acting under the conviction that as Secretary of Agriculture his long experience in dealing with Governmental affairs would enable him to improve the public service.

Promotions which Moore had given the higher officials since September, 1912, were all cancelled, and the salaries in effect prior to that time were put back into effect without prejudice. Officials who had been suspended were reinstated. Chief Moore and the Chief Clerk of the Weather Bureau were dismissed.

CHAPTER XXVIII.

HURRICANE LISTENING POSTS.

TROPICAL cyclones, with their hurricane winds and destructive storm tides, have always been of great concern, to ocean shipping and to residents of coastal areas. The United Fruit Company had equipped their ships with wireless telegraphy, and had established radio telegraph stations at Swan Island in the Caribbean Sea and at Cape San Antonio, Cuba. The Weather Bureau needed all the weather reports from the areas traversed by tropical cyclones that could be obtained. The United Fruit Company cooperated with the Weather Bureau in collecting reports of the weather during these disturbances, and arrangements were made to locate weather observing stations at their wireless stations and to use their operators as observers.

The establishment of observing stations at Swan Island and Cape San Antonio was decided upon in April, 1913, and the assignment of putting the stations into operation was given to me. The equipment for these stations consisting of barometers, thermometers and other instruments, was assembled at New Orleans. The journey to Swan Island was made on a United Fruit Company steamer which, dropped me and the instrumental equipment into a small motor boat about a quarter of a mile off shore from the Island. This boat carried me to the Island. The national ownership of the Island appeared to be in question. However, the Stars and Stripes floated freely from a mast near the landing place, a sight always pleasant to behold.

Swan Island of volcanic origin was small and had no permanent inhabitants except sea fowl which through the ages had deposited there a wealth of fertilizer. The story runs that a Mr. Adams of Baltimore, Maryland, who operated a fleet of sailing vessels engaged in collecting fertilizer, discovered Swan Island sometime around 1880. He claimed it in the name of the United States and hoisted the Stars and Stripes. The United States Government took no notice of Mr. Adams' claim. However, this did not deter Adams. When it was found that he was hauling boat loads of fertilizer from the Island the government of Spanish Honduras laid claim to ownership. The Honduran navy chased Adams away, but he was not going to be deprived of his find. Adams armed his boats with cannon, and went back to Swan Island prepared to back his claim with shot and shell. The Honduran Navy attacked, but Adams won the battle and held the Island. Besides gathering fertilizer he planted a fine orchard of cocoanuts, which produced a large yield. The Island had a natural growth of tropical fruits. In fact it was a "Garden of Eden", but without an Eve, a most important omission. No woman had been allowed on the Island up to the time I was there. Laborers were brought in from Jamaica and other neighboring British Islands. They were kept for six months, then another batch was brought in for six months, and so on.

When I arrived several men from New Orleans were constructing buildings for the United Fruit Company. The instrumental equipment was set up, and the wireless operators were instructed in the art of taking observations and coding weather reports. I was now ready to proceed to Cape San Antonio, Cuba. The journey was a long and arduous one. The United Fruit Company sent a steamer by Swan Island to pick me up and start me on the way. To reach the post in Cuba it was necessary for me to go the "wrong way". I had to go first to Puerto Lemon, Costa Rica, and there I was to transfer to another ship bound for Havana, from which place I would go to Cape San Antonio by an undetermined route. We were delayed en route to Puerto Lemon by the treacherous winds of a tropical cyclone, and, as we approached the port, met the ship on which I was to sail well out on its way to Havana. On arrival at Puerto Lemon I was transferred to a ship sailing for Colon, at which place I was to be transferred to another ship bound for Havana.

After spending a few days in Colon looking over the great Panama Canal I boarded the steamer for Havana and reached there without any mishaps. Serious difficulties were encountered in reaching Cape San Antonio. No regular transportation route was in existence. The American Consul was consulted as to means of getting there. He told me that it would be hazardous to attempt the trip over the mountains with a donkey train, and he could not suggest any route of travel. This left me "up in the air" for my time was limited and some way to make the journey had to be found quickly. I had the good fortune to meet an American mining engineer, operating in Cuba, who had friends in all parts of the Island. He told me that Cape San Antonio was almost inaccessible; that it was a hide-out for men who had been guilty of wrong doing. To escape punishment, they had retired from public view, and joined the colony of charcoal burners on the Cape. My engineer friend assured me that the charcoal burners were dependable, and that when they formed a friendship for any one they would go the limit in lending assistance. He contacted his acquaintances around the coast to the Cape and requested them to be on the lookout for me and render such assistance en route as I might need. He assured me that this would make the journey safe from the perils which might otherwise confront me.

My friend arranged passage on a small steamer, called the "Vapour", which would take me to within about 75 miles of Cape San Antonio. There a friend of his arranged for me to charter a sail boat with a cabin to continue the remainder of the journey.

At last I was ready for the trip which promised some excitement. The "Vapour" had no cabin accommodations. There was one large room in the center of the boat, leaving the front and rear portions of the deck for the storage of freight and animals. Passengers frequently took their cattle, donkeys, chickens, etc., on the boat with them. The large room was used for a dining room as well as a sleeping room. Evidently I was considered a distinguished passenger for when night came a table was fixed up as a bed for my use in one corner of the room. The other passengers, men, women, and children piled down on the floor and slept in the other corners of the room. After a good night's sleep I was awakened in the morning by something soft and warm caressing the side of my face. I turned over to see what it was, and beheld a fine bull, his head sticking through a small window. He evidently had mistaken my beard for hay, and was preparing to help himself.

We arrived at the termination of the "Vapour" line, and a friend of the mining engineer, a tall Texan who had remained in Cuba after the Spanish-American war, met me at the boat landing. He said that he had chartered a sail boat to take me to the Cape, but the boatman was moving a family down the coast and it would be four days before the boat would be available. I could not wait, so he made arrangements for me to go with the mail carrier on a small sailing boat leaving at 2:00 o'clock in the morning, for the Light House on the Cape, to reach there the same day if the wind was favorable. When I boarded the boat the first thing I saw was the menu for the day, a small red rooster tied to the mast, a piece of fat side from a wild boar, and some red beans. There was coffee of course. Under favorable winds we made good progress, and hauled up a few hundred yards off the beach in front of the wireless station. I took off my shoes, rolled up my pants and waded ashore. This was on a Tuesday. The men at the wireless station expected me, for I had radioed them from the ship as I passed by in sight of them several days before. The instrumental equipment was put in position and the manager of the wireless station was instructed in the duties of observer. Saturday morning found me ready to start for Havana. My friend, the mining engineer, had arranged for a charcoal schooner to pick me up Saturday afternoon and take me to Le Fe en route to Havana. Saturday about 6 p. m. the schooner came in sight and the boatman sent his one assistant ashore to get me and my luggage. The boat had one small cabin underneath the deck. Its overall length was not more than twenty feet, with one mast to carry the sail. The boatman promised to land me in La Fe Sunday morning. I went into the cabin and was soon sound asleep. When I awoke Sunday morning I looked out to see how near we were to La Fe. Imagine my surprise when I saw we were still in sight of the wireless towers at Cape San Antonio. The deck of the boat was crowded with charcoal burners, including one woman and baby, which the captain had picked up during the night. These men looked like they might have been pirates. Their long black mustaches and shaggy hair gave them a vicious appearance, and long knives hung from their belts. The rain was pouring down and the wind was blustery. The captain of the schooner said that it had been too stormy to make the trip during the night, but that he would have me in La Fe Monday morning. I insisted that the woman with the baby take the cabin. The men all demurred, but I turned the cabin over to the woman and baby so that they would be out of the rain. I took my place on the deck with the charcoal burners. It rained throughout the day, and we made slow progress. With rain coat and oil cloth with which I had kept the instruments dry, I was able to keep myself fairly comfortable. The wind gradually died down during the night, and soon after midnight the sails showed that there was no wind to carry us on our journey. The Captain said there would not be any wind before sunrise. It was now about 1:00 a. m. Monday, and he was to have landed me in La Fe by 7:00 a. m. He put two of the toughest looking men in the lot into a skiff with two sets of oars, and moved me into the skiff with my belongings, remarking, "They take you to La Fe". He pushed us off with his blessing. Here I was in the hands of two men who looked the part of pirates, no land visible, not even a light, and worse, not even a star to guide us. Nothing could be seen, there was the dark heavens above me, and the briny deep surrounding me with this small skiff my only haven. I repeated in my mind these lines from the Psalmist: "The Lord is my Shepherd I shall not want, He maketh me to lie down in green pastures, He leadeth me by the still waters". I called to mind the advice "Trust in the Lord but keep your powder dry". I handed each of the men two silver dollars and said, "Two more at La Fe". They then handled the oars with swiftness and force. They made their way on one of the darkest nights direct to La Fe, and landed me there at 7:00 a. m. I was so pleased with their accomplishment that I gave each of them four dollars instead of the two I had promised them. They went back to their skiff and rowed off to find the schooner they had left becalmed at sea.

La Fe was only a coastal village, but I soon located a **bolante** (a four-wheel covered spring wagon with two seats) which was waiting for me. The bolante was drawn by two mules, which the driver kept moving at a good pace—in fact they traveled so rapidly over the rough road that my head was against the top about as much of the time as my feet were on the floor of the vehicle. We arrived at San Juan where I was to take the train for Havana. The driver was paid his fee, but I noticed that he was at my heels all the time. When I entered the restaurant to get my dinner he seated himself at the table with me. We each ordered our dinners. After we had finished eating I paid for both dinners, and the driver left and I saw no more of him. When the train for Havana pulled into the station it was a good sight to behold. However, the mining engineer's bandit friends had taken good care of me. When I boarded the train I noticed two soldiers at each entrance to the coaches with their guns ready for quick action. The passenger in the seat with me was asked the reason for the soldiers being there. He said robbers attacked the train every two or three days, and the soldiers were there to prevent robberies. They did not attack this train, and we reached Havana on schedule time. A United Fruit ship was in harbor, and I boarded it, arriving in New Orleans on June 30, 1913. The hurricane listening posts were in operation, and the daily and special reports of observations were being received at New Orleans.

CHAPTER XXIX.

PROFESSOR MARVIN APPOINTED CHIEF OF THE WEATHER BUREAU.

HIS attainment as a scientist was the principal consideration in the selection of a Chief of the Weather Bureau to succeed Willis L. Moore. Prof. Charles F. Marvin, the man appointed by President Wilson to that important position had rendered distinguished service in several branches of science. He gave up a professorship in the University of Ohio to become Junior Professor of Meteorology in the Weather Service in 1884. He had contributed more than any other one man to the success of the Weather Bureau. As a student of mechanics he specialized in the development of accurate meteorological instruments. The tables used by the Bureau for determining the moisture in the air are based on experiments which he conducted. He invented instruments for measuring and recording rainfall, snowfall, sunshine, and atmospheric pressure. He made special studies of anemometers, and developed improvements in instruments for recording wind velocity and direction. He studied and published contributions on the use of kites in making meteorological observations in the air above the earth's surface. He was an all round meteorologist, always ready to aid others in research work. He had none of that personal jealousy exhibited by many other officials.

Prof. Marvin was too big a man to permit himself to be drawn into the cliques which tried to use the Bureau for their personal aggrandizement. He dealt fairly and honestly with every one with whom he was associated. He expected every man to do his duty and he asked nothing more. I had known him, through a personal friendship with his Chief Clerk, from the time he entered the Weather Service. He was especially fitted to clear away the debris left from the explosion which Moore had set off. The Bureau needed a Chief who was not only capable, but who would give every one a square deal and thus instill confidence in the personnel. He made the rounds of several of the more important stations soon after he was appointed Chief. When in New Orleans I showed him the letter in which Moore had personally directed me to limit the forecasts of flood crests in the Mississippi River to ten days in advance of their expected occurrence. He read the letter and said: "Cline, you can disregard this letter and issue warnings for any period which you think will best serve public interests". He soon brought the service to a high degree of efficiency. He would not permit political influence to enter into the affairs of the Weather Bureau. When an official felt that he had been unjustly treated he could go directly to his Chief where he could get a hearing.

In a previous chapter I told of the frequent assignment of assistants to my station whose conduct had been such that the officials under whom they had been serving could not manage them. Moore, during the last years of his administration had permitted the Chief Clerk's Office to make my station the dumping ground for incorrigibles. This was done in order to force me to discipline them and to try to make it appear that I was unreasonably harsh in dealing with my assistants. When Prof. Marvin became Chief there were two men on my station who had been transferred to New Orleans because they had caused trouble on other stations. These men soon became insubordinate and neglectful of their duties in so flagrant a manner that I was compelled to recommend that they be disciplined. The reply I received from Prof. Marvin confirmed my suspicion that undesirable assistants had been sent to my station through the contrivance of a certain clique in the Central Office in order to "get" me. Prof. Marvin wrote me that representations had been made to him by the Chief Clerk's Office (William Weber was not a party to this; he was one of the fairest and cleanest men in the Bureau) that I was unreasonably harsh with my assistants and was continually having trouble with them, and that on this account he hesitated to carry out my recommendations, asking me to deal less harshly with these men. Soon after this he called me into personal conference. I showed him the records of the men in question at their previous stations, and also gave him a list of the records of other men with whom I had trouble. It was for such purpose that I had kept the records of the trouble makers. Prof. Marvin checked on the records of all the men I had cited and found the facts as I had stated them. He sustained me in my recommendations and disciplined the men in question. He told me that in the representations made to him the previous records of the men had been kept from him. He saw in this an effort to discredit me, and afterwards he looked carefully into any recommendations affecting my station.

When the United States entered World War I, experienced men were taken off my station and I was instructed to employ temporary assistants in their stead. These instructions were carried out without making any complaint because I was anxious to contribute my part to the war effort. In December, 1917, I visited the Central Office and went to the Chief Clerk's Office. The Official in Charge of the Weather Bureau Office at Corpus Christi, Texas, had been a reserve officer in the German Navy, and it was important that he should be sent to a station where he could be watched. Mr. Calvert, the Chief Clerk, remarked that they were thinking of transferring W. F. McDonald from my station to Corpus Christi. I objected, and told him that if they would promote McDonald to First Assistant and leave R. A. Dyke as assistant forecaster they could take all the other men and I would carry on the station work with temporary assistants. Just then Prof. Marvin sent for me to come to his office. I told him of the conversation with the Chief Clerk. Prof. Marvin rang for the Chief Clerk, and as he entered the Chief said: "I told you not to disturb the New Orleans station again." Prof. Marvin ordered that my recommendation about Messrs. McDonald and Dyke be carried out. My station was not used as a dumping ground for suspicious persons or incompetents as long as Prof. Marvin remained Chief of the Weather Bureau.

Prof. Marvin was Knighted by the King of Norway for his efficient work in furnishing weather information to Roald Amundsen while on polar explorations. Marvin instituted the practice of giving pioneer transatlantic aviators special weather reports and data to help them on their way. He invented the clinometer, an instrument for use in determining the exact height of clouds. He was the author of many technical papers on meteorological subjects. Prof. Marvin was a member of the National Advisory Committee for aeronautics, 1915-1934, and of the National Research Council in 1917. He was secretary on Meteorology of the International Geophysical Union organized at Brussels in 1919.

Prof. Marvin called for volunteers for meteorological service in the Fire and Flame Division in France during World War No. 1. When I heard of the call I volunteered for the service. He told me that my services were needed in New Orleans, and that a man who could not do my work here could do the work in France.

CHAPTER XXX.
THE 1915 TROPICAL CYCLONE AT NEW ORLEANS.

HURRICANE listening posts which had been established at Swan Island and Cape San Antonio proved their value in the tropical cyclone which made its appearance in the Caribbean Sea on September 22, 1915. This was the largest tropical cyclone, and was attended by the highest wind velocities of record, to reach the Gulf coast in the history of the Weather Bureau. The cyclone passed through the Yucatan Channel during the night of September 27th, bearing directly on the middle Gulf coast.

I had learned enough about the action of storm tides to enable me to give advance warning to the regions where they would occur so that people could be moved out of the danger zones before the arrival of the tides. This cyclone, before it had passed through the Yucatan Channel was sending its message through that channel to the Gulf Coast in the form of long swells which commenced building up a storm tide on the coast. At 8 a. m. of the 26th there was a storm tide on the middle Gulf coast of 0.8 of a foot. At 8 a. m. on the 27th the water had commenced rising at Fort Morgan and was 1 foot above the predicted tide on the Gulf coast from Burrwood to Galveston; at 8 p. m. the excess was 1.1. feet. At 8 a. m. of the 28th, there was 1.5 feet of water in excess of the predicted tide at Burrwood and Galveston and 0.8 of a foot at Fort Morgan. At 8 p. m. of the 28th when the center of the cyclone was about 250 miles distant from the Gulf coast the storm tide was 1.7 feet at Burrwood, La., (at the mouth of the Mississippi River), and there was no storm tide at Galveston, Texas. The rising storm tide at Burrwood indicated that the center of the cyclone was moving towards a point not far west of the mouth of the Mississippi River, and that all low lands along the east Louisiana, Mississippi and Alabama coasts would be flooded to dangerous depths extending well inland. At 2 a. m. of the 29th the storm tide was 2.7 feet at Burrwood and 2 feet at Fort Morgan, Alabama, and was spreading over the low lands along the coast at an alarming rate. Six hours later at 8 a. m. of the 29th, Burrwood had a storm tide of 3.7 feet, Fort Morgan, Alabama, 2.5 feet. The storm tides indicated that the center of the storm would move inland over Barataria Bay.

Southeastern Louisiana experienced the worst part of the storm. It was covered with a storm tide of 10 to 13 feet, which extended up into Lake Pontchartrain around New Orleans. On the Mississippi coast the storm tide was 9 to 11 feet, and on the coast of Alabama and Northwest Florida 4 to 9 feet. Pensacola, 175 miles to the right of the center of the cyclone, had a storm tide of 4 feet; Mobile, 150 miles to the right of the center, a storm tide of 7.2 feet; Gulfport, Mississippi, 80 miles to the right of the center, a storm tide of 9.6 feet; and Bay Saint Louis, Mississippi, 65 miles to the right of the center, a storm tide of 11.8 feet. The highest tide was 13 feet at the west end of Lake Pontchartrain. There was no storm tide to the left of the point where the center of the cyclone moved inland. Morgan City, La., 25 miles to the left of the path of the center of the cyclone had an unusually low tide, 0.4 foot below mean gulf level.

The cyclone moved northward, curving towards the east with its center passing, inland just west of Barataria Bay, thence northward over Tulane University and between Lake Maurepas and Lake Pontchartrain. Burwood had a maximum wind velocity for five minutes of 140 miles per hour, and a sustained velocity of 107 and 108 miles per hour for two consecutive hours.*

The knowledge which we had acquired in the study of storm tides enabled us to foresee the dangerous conditions described in the above paragraph. We issued extraordinary warnings with definite advice to the people in the threatened areas, indicating where the storm tide and hurricane winds would be dangerous to life and property. People in certain sections were warned that their lives were in great danger.

---

* See TROPICAL CYCLONES, Cline, The MacMillan Company, New York, pages 110-129, and 259-263.

The telegraph and telephone were used to the fullest extent; and localities which could not be reached in this way were reached by special messenger at Government expense. Motor boats, automobiles and every means of conveyance, even to the horse cart were used.

The entire fleet of vessels of the Jahncke service was mobilized and sent into threatened sections, bringing people to safety. Many other owners of pleasure and commercial boats helped in the same manner.

Schools were advised early on the morning of the 29th to close for the day, and to send the children home to their parents. Merchants were told to close and barricade their stores, and send their employees home. The police and Fire Departments were instructed to warn people to stay off the streets to avoid being killed or injured by wreckage which would be driven through the streets by high winds like missiles shot from a cannon. The railroads were told to stop their trains at Slidell—that trains endeavoring to cross the bridge to New Orleans in the afternoon would be blown off the bridge into the Lake Pontchartrain by the hurricane winds. All ocean shipping was advised to remain in port. Our instructions were heeded and the loss of life, when the extent of the area covered and fury of the storm is considered, was very small.

There was only one notable instance where the warnings were not heeded. A man named Manuel and his wife were keepers of a club at the Rigolets. Some 25 people lived near the club house. Manuel was called about 9 a. m. on September 29th over the long distance telephone by the Weather Bureau Office, warned of his danger and told to have everybody come to New Orleans on the next train—their last opportunity to escape with their lives. Manuel said, "That train will not stop on a flag signal for us". He was told to put a crosstie on the track and put up a danger signal. Manuel replied, "They will put me in jail". He was told, "You would be better off in jail than where you are now, and for God's sake stop that train at all hazards and come to New Orleans". Manuel stopped the train, but the people were not there to get aboard the train. The rising storm tide was jeopardizing the lives of the passengers on the train and it could not wait for Manuel to go and collect the people from their homes. Manuel returned to his companions, and when the storm had passed and the tide had receded, the sun beamed down on the lifeless bodies of Manuel and 25 of his friends scattered over the marshes.

Exactly at 12 midnight of the 29-30th with the wind still blowing 30 miles per hour, Mr. Albert R. Israel of the Associated Press braved the elements and reached the Weather Bureau Office. He informed me that a wireless telegraph station had been opened in a ship on the river front. He was asked if he could get a message through to Washington for me. He said, "Write a 150-word message, and I will see that it goes to headquarters." Mr. E. D. Coberly, my First Assistant, went with him and filed the message at the wireless station on a United Fruit Company ship. Thus before the storm was over the Chief of the Weather Bureau had a concise report on the situation and the essential facts about the storm.

Nearly every building in New Orleans was damaged to a greater or less extent. Notwithstanding the extent and severity of the storm only 275 persons, some of whom were at sea, lost their lives in the thickly populated coastal area. This was a comparatively small loss of life, and proved the great value of advance definite information concerning the approach of a hurricane. Previous tropical cyclones of less extent and severity had drowned as many as 2,000 persons. The property damage resulting from the storm tide and hurricane winds of the 1915 hurricane was estimated at $13,000,000.

A steamer, the Marijwine, with its crew and passengers, went down during the storm in the Yucatan Channel. Strange as it may appear none of the bodies of those lost were found, and not a piece of the ship was ever seen. The boat probably capsized and was covered with drifting sands carried by cyclone currents.

The New Orleans Times-Picayune of September 30, 1915, commenting on my work in the above hurrciane said:

> The intensity of the storm, while it did considerable damage in New Orleans and vicinity, proved the worthiness of Dr. I. M. Cline, forecaster of the United States

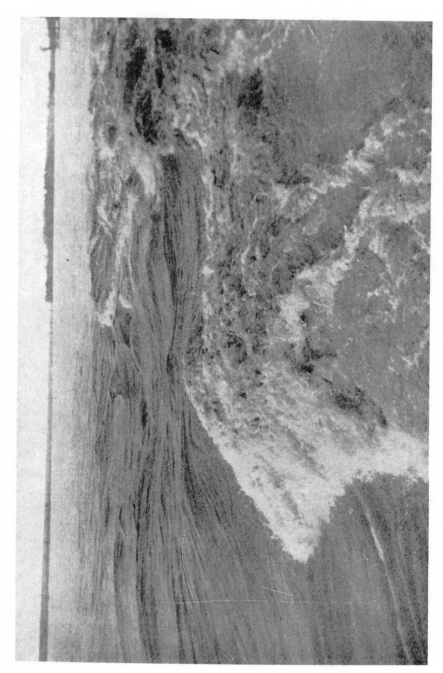

Mississippi River Crevasse. Rush of water through a break in the levee.

Weather Bureau. Never before, perhaps in the history of the Weather Bureau, have such general warnings been disseminated as were sent out by the local bureau in reference to the disturbance that passed over New Orleans Wednesday evening * * * * There may have been much loss of life along the coast but such a catastrophe can not be charged to the Weather Bureau, for the warnings of the approach of the hurricane were sent broadcast before the high winds ever reached the Louisiana coast.

The New Orleans Item on October 13, 1915, wrote:

About 20 years ago a West Indian hurricane, far lighter in force and stress than the recent storm, struck the Gulf coast. Over 2,000 lives were lost and many millions in property.

Ten days ago another West Indian hurricane came with tremendously increased intensity. But the loss in life in all the vast stretch of marsh and bayou and sea line is only 275. The property damage is infinitely less.

There is one specific reason for this difference in results: Increased efficiency in the Weather Bureau and an increased and extended service rendered possible by enlarged personnel and extended range of observations.
(See U. S. Monthly Weather Review, September, 1915, pages 456-466.)

CHAPTER XXXI.

CLIMATIC CHANGES AT NEW ORLEANS

WASHINGTON was the place for the meeting of the Second Pan-American Scientific Congress held December 27, 1915-January 8, 1916. Prof. C. F. Marvin, Chief of the U. S. Weather Bureau, instructed me to attend that Congress and read a paper on "Temperature Conditions at New Orleans as Influenced by Subsurface Drainage". I could not find any previous study which showed that the climate of a locality had been influenced by artificial changes in its surroundings. New Orleans is probably the only place in the world where artificial changes in its natural physical conditions have been extensive enough to change the temperature of the overlying atmosphere.

The natural physical conditions as they existed in and about New Orleans may be summed up as follows. Until recently portions of the waters of the Mississippi River found their way to the Gulf of Mexico to the East of New Orleans through Manchac Bayou, Lake Maurepas and Lake Pontchartrain, by way of the Rigolets into Lake Borgne and into Mississippi Sound. On the West, water from the Mississippi found its way to the Gulf through Atchafalaya Bay by way of the River of the same name. There were many other smaller bayous traversing both the east and west delta lands which carried off part of the mighty river's overflow to discharge it into the Gulf of Mexico. The extension of the levee system has gradually closed all these outlets to the Gulf, except those through the Atchafalaya, and the Mississippi River proper. The Mississippi passes through and almost around New Orleans, hence the name of **The Crescent City**. The river at New Orleans in places is more than 100 feet deep and more than a half mile wide. From the City it winds in serpentine manner to the Gulf of Mexico, covering about 100 miles in its course. The closing of natural outlets, while they reduced the area of surface water to some extent, forced the water into secondary channels, which are wide and in many places of considerable depth, and the water in large quantities backs up through these into Lake Pontchartrain and Lake Maurepas to the north of New Orleans, and up through Barataria Bay, Little Lake, Lake Salvadore, and Lake des Allemands on the west. Thus is formed an almost complete inclosure of water of considerable area around the city, which spreads out over the marshes, and frequently, with the rise of the tide and during the prevalence of easterly winds reaches a depth of 3 to 4 feet.

The coast line of the delta region is very irregular, indented by numerous bays, and is cut up by thousands of small lakes and Bayous into a labyrinth of peninsulas and islands. The general shape of the coast line of this area is the arc of a circle convex to the Gulf, the radius of which is about 65 miles. The greater part of the land is marsh, the fertile alluvial land interspersed being the only portion not subject to overflow by tidewater.

Surrounded by these areas of water with the marsh land frequently covered with water, the elevation of a considerable area of the City of New Orleans is only a few feet above tidewater. In the past storm and flood waters were carried off over the surface of the ground, and the ground water under normal conditions, was level with the surface of the soil, all of which combined to produce insular climatic conditions.

So that we may better understand this question we will notice briefly the influences of physical conditions on the temperature of a locality. The variations in temperature over water and over land, and the differences in the effects resulting from radiation over water and over the ground, and the gain and loss of heat by radiation are the chief factors in determining the temperature of a locality. Dry land is heated much more rapidly by solar radiation than water, the specific heat of water being much greater than that of dry land. Equal quantities of heat acting on equal areas of land and water increase the temperature of the land nearly twice as much as they do the temperature of the water. This does not take into consideration the fact that an increase in the temperature of the water is further lessened to a material degree by evaporation from the water surface, thereby rendering latent a large amount of the heat received from solar radiation. Solar radiation penetrates the water to a depth more than ten times greater than it does the ground, yet at the same time the water surface is heated to a much less degree than the surface of the ground. Water in turn gives off its heat to the atmosphere by radiation and conduction much more slowly than the ground, so that we have a slow heating up during the day and a slow cooling off during the night in the atmosphere overlying bodies of water as compared with a rapid heating up during the day and a rapid cooling off during the night over the land. Water, when of considerable depth, stores up a much greater amount of heat during the summer and gives it off much more slowly during the winter than land, and we have a lower day and higher night temperatures and lower summer maximum and higher winter minimum temperatures over bodies of water than we find over bodies of land in the same latitudes. It is not necessary that the ground be covered with water to obtain these results in some degree, because the very wet soils warm up more slowly in the summer than dry soils, both because of the high specific heat of water contained in the soils and the effects of cooling by evaporation.

In New Orleans prior to 1900 the water in the ground was almost level with the surface of the earth. Water continually flowed through open drains on each side of the streets and in many places very sluggishly. All storm water was carried off over the surface of the ground. Besides displeasing to the eye this condition was unsanitary and favored the breeding of disease bearing mosquitoes and other water insects. Such conditions, while favoring an equable climate, were not conducive to healthfulness. To overcome these objectionable conditions plans were made to drain the city. The operation of subsurface drainage was commenced in 1900, and an immediate result was the lowering of the water in the drainage canals 8 to 10 feet below what it had been previous to that time. This effected a much more rapid transfer of storm and other water from the city through the drainage system into Lake Pontchartrain.

The operation of sanitary sewers was commenced in 1903. Prior to that time all sewer water was discharged into wells, one or more to each residence plot, walled up and sealed over so that the water from the sewer was distributed through the soil by diffusion and gravity and then carried to the surface with the level of the ground water then existing. The results obtained by the new sewerage system may be summed up as follows: No water flows in the open gutters on the streets as was the case prior to 1900; storm waters and the millions of gallons of water discharged daily into the sewer wells to spread out through the soil and maintain a high ground water level, now go out through the drainage and sewerage systems; and the ground water, instead of being level with the surface of the ground, is 6 to 8 feet below that surface.

On nearly all sides of New Orleans large areas of marsh lands which were covered the greater part of the year with tidewater have been drained, thereby reducing the area covered with water to a much less extent than formerly.

It has been noted in a general way the influences of water and land on radiation in determining the temperature of a locality, but there were no records made previous to the observations at New Orleans to indicate to what extent the removal of the surface water from a considerable area will influence its temperature conditions. Under ideal conditions we should find a decided change in temperature where a large area has been changed from a water surface or a ground surface saturated with water to comparatively dry ground. But we do not find ideal conditions for the study of the effects of solar and terrestrial radiation on atmospheric temperatures. They are particularly far from ideal over a city. Cloudiness retards cooling from terrestrial radiation at night to a greater extent than it retards warming up from solar radiation during the day. Over large cities smoke and dust particles suspended in the atmosphere, hundreds of thousands of dust motes being found in a cubic centimeter of air, also retard cooling from radiation at night. The inflow and mixture of air from surrounding areas also plays a part, the extent and influences of such sources depending on the direction and velocity of the wind.

Evaporation being an important factor in retarding a rise in temperature from the effects of solar radiation on a water surface, the influence of shallow water and a wet soil, such as prevailed in New Orleans prior to the installation of subsurface drainage, would be much more pronounced in their effects on day temperatures than on night temperatures and on summer temperatures than on winter temperatures. Shallow running water and a saturated soil, while preventing a material rise in day temperatures as the result of radiation from the sun during warm periods, would not store up heat to materially influence night or winter temperatures.

Comparison of temperatures for the 15 years, 1885 to 1899, inclusive, which covers the period just prior to the installation of subsurface drainage, with the temperatures for the 15 years, 1900 to 1914, inclusive, during the operation of subsurface drainage with the thermometers in the same exposure during both periods, brings out some interesting facts. The temperature at New Orleans had never reached 100 degrees prior to 1900, but in 1901, a temperature of 100 degrees was recorded and temperatures of 100 degrees or higher were recorded seven times in the 15 years following 1900. Temperatures of 95 degrees or higher were recorded during the fifteen years ended with 1899 on 35 days, while in the 15 years 1900 to 1914, inclusive temperatures of 95 degrees or above were recorded on 74 days—an increase of 112% over the number of days with such temperatures in the 15 years just prior to the installation of subsurface drainage. The highest temperatures occurred after physical conditions at New Orleans had changed. The average daily temperatures were higher also. These records thus furnish the best material for use in determining the effect of such changes on the temperature of a locality.*

---

* The complete study of this subject will be found in the Proceedings of the Second Pan-American Congress, 1915-16, Washington, D. C., Volume II, pages 491-496.

CHAPTER XXXII.

PORTRAITS OF AMERICANS
By AMERICAN PAINTERS

PORTRAITS by early American painters who founded American art, have always interested me. I decided in 1903 to utilize my recreation and vacation time buying and restoring old paintings as a hobby. The preservation of early American portraits for future generations appeared to me as a promising and worthwhile venture. There were three reasons why I decided to undertake this work. In the first place I was confident that in time there would be a demand for such portraits by museums and collectors; in the second place I was confident that my cash investment and the time spent in collecting and restoring the portraits would bring good financial returns; and thirdly, the preservation of the portraits of pioneer families and the work of the artists would be a contribution to American history. The majority of old portraits had been neglected and were in bad condition. They were fast going to ruin because there had been no interest in their preservation. Louisiana was a rich and an unexplored portrait field. I was warned by many that I was investing in something with no future, and that my time and investment would be a loss. My father and Dr. Ellsworth Woodward, Director of the Newcomb Art School, were the only ones who gave me a word of encouragement. The very fact that my venture was cried down by artists and others proved to my advantage. I was pioneering in a field which is now very popular. American portraits which I owned at the time of my art sale in 1907, were reserved as a nucleus around which to build a collection. When a good portrait appeared on Royal Street in the antique stores, the second hand stores, or at auction, I was there to buy. Families often died out leaving their effects to be sold and the proceeds distributed among distant relatives who were not interested in the portraits. Often the paintings had been badly torn, and needed to be relined, the holes filled in and the colors matched. No retouching was done, for that not only destroys the originality of the artist's work, but changes the likeness of the person represented by the portrait. Portraits by such painters as Gilbert Stuart, Thomas Sully, John Wesley Jarvis, Matthew Harris Jouett, William Jewett, Eliab Metcalf, Samuel L. Waldo, Samuel F. B. Morse (inventor of the telegraph), Chester Harding, Rembrant Peale, and many others were purchased, restored and saved from ruin. There was a thrill in bringing back to life a fine painting which some artist years before had taken pride in executing for the delight of the living and to be handed down to posterity.

Many of the portrait painters did not sign their paintings. This was particularly true of Stuart, who claimed that his paintings were signed all over; by which he meant that his style was so individual that his pictures could always be recognized. Dunlap in his "History of the Arts of Design" stated that he had seen Stuart painting the portraits of certain persons. Portraits of these people were in my collection. I was anxious to authenticate the artist who painted these portraits. Charles Henry Hart of New York City was recommended to me as an authority on American portraiture and especially the work of Gilbert Stuart. I wrote Mr. Hart and he requested me to send him photographs of the portraits which I thought were painted by Gilbert Stuart. He wrote me that he was sorry to have to tell me that my portraits were not by Gilbert Stuart but by Ralph Earl. I questioned his decision and told him that Dunlap stated in his book HISTORY OF THE ARTS OF DESIGN that he saw Stuart painting those people. Hart wrote back that Dunlap was very unreliable and that his book was full of errors. I wrote Hart: "Dunlap was a man of such high standing in his day that he was elected President of the National Academy of Design, and what he said that he saw will stand when what you say is rotten and forgotten". I will leave the reader to judge the honesty of Hart's statement when he reads what is brought out later in this chapter. This was in 1913 and no further effort was made to identify the artist who painted the portraits. My intention was to sell the collection of portraits at some future time and leave it to the purchaser to find out who painted the portraits.

About 6 p. m. one day in the latter part of December, 1917, the door bell at my residence rang and I answered. A gentleman handed me his card—"Charles X Harris, Artist, New York, N. Y."—with the remark that he was interested in early American portraits. He had been told that I had a collection of such portraits, and asked if he might be permitted to see them. I ushered him into a room where there were some eighty fine American portraits covering the walls from ceiling to floor. As he entered the room he lost his poise and said audibly: "Here is what I have been looking for." I made a mental note of that and resolved that if he got them he would pay for them. He had aroused my interest, and I asked him to have supper with me. During the supper neither of us mentioned the paintings. I was on forecast duty and had to be at the Weather Bureau Office at 7:30 p. m. Mr. Harris had accommodations on the Louisville and Nashville Railroad to New York, and he went down on the street car with me. He asked me if I ever sold any of my portraits. I told him that I was a collector and that I had nothing for sale. Immediately on his arrival in New York he wrote me stating that he had a client who was much interested in my paintings. He said that unless I contemplated giving them to the National Gallery or some other gallery he would be interested in purchasing them. I wrote him that while I felt that the portraits should belong to the National Gallery I was not financially able to donate them. I told him that I realized that the portraits located in a frame building in New Orleans might be destroyed by fire and their destruction would be a great loss to the nation; on this account and in order to get the portraits where they would be better protected I would sell them for a reasonable consideration. Mr. Harris did not wait to write, but telegraphed, asking my price for the collection. I wired him that the price was $35,000.00. He telegraphed back: "Price too much, shall we come on for further negotiations?" I wired him to come. This was on Friday, early in January, 1918. The following Monday morning Mr. Harris and another man, the financial agent of Thomas B. Clark, showed up at my residence, saying: "We have come to talk turkey, what is your lowest price?" I replied: $30,000.00." Harris said that was too much and that they could not pay that price. I asked them what they expected to pay and Harris handed me a letter of credit from a New York bank on the Whitney National Bank of New Orleans for $25,000.00. That was more money than I had ever seen at one time, and I sold them the collection for that sum.

I inquired of Harris about how he came to visit me in December. He said that Charles Henry Hart had told Thomas B. Clark, the portrait collector, that he had located certain Stuart and other paintings they were looking for in the South but would not divulge their location. Hart had promised them he would get the paintings for them, but five years passed and he had not brought the paintings. Harris said that Clark was tired of waiting on Hart and had sent him in search of the paintings without consulting Hart. Harris said that he had visited every town in the South from the Atlantic Coast to the Rio Grande. He had saved New Orleans for his last city to visit, and at 4 p. m. the day he called on me had given up all hope of finding the paintings and had bought his railroad tickets back to New York. He then met some one who told him that Dr. Cline had a collection of portraits. He found what he was looking for just as he was about to give up the search.

I paid off all my debts, and started another collection. I purchased $18,000.00 of War Bonds and had them registered in my name.

Andrew Mellon in the course of time became owner of the paintings, and several of them are now in the Mellon collection in the National Gallery at Washington, now the property of the United States. My recreation time having been used to good advantage both for myself and the people of the nation I was ready for another venture.

CHAPTER XXXIII.

THE CORPUS CHRISTI HURRICANE OF SEPTEMBER, 1919.

TROPICAL cyclones seldom travel westward over the Gulf of Mexico, but when they have followed such a course, they were of vast extent and unusual severity. An abnormal movement of the upper air over the Gulf of Mexico prevailed during the first half of September, 1919. There was a steady east to west movement of the upper air, and this carried the cyclone westward and prevented it from recurving to the north and then to the east which is the normal movement for cyclones in this region. Such a phenomenon called for unusual alertness on the part of the forecaster. Here he was called upon to use every bit of information which might give him a clue to the direction in which the cyclone would travel.

The cyclone passed through the Florida Straits during the night of the 9th of September, with the lowest barometer at Key West, Florida, reading 28.81 at midnight. During the 10th, 11th, and 12th, the cyclone traveled very slowly towards the northwest, averaging about 7 miles per hour. The forces which normally control the direction of travel of a tropical cyclone at this time of the year were having an aerial combat with abnormality in the upper air. During the afternoon of the 12th the abnormal movement of the upper air current from east to west took the ascendency, and overcoming the normal tendency to recurve, carried the cyclone westward with a rate of travel of 14 miles per hour, which rate it maintained until it moved inland. The center of the cyclone on the 12th was about 400 miles south of the middle coast of Louisiana, and the diameter of the cyclone was about 800 miles. Barometer readings at New Orleans, Burrwood, and Galveston conveyed very little information relative to the extent and intensity of the cyclone and the direction in which it would travel. The lowest barometer recorded at New Orleans was 29.72 at 7 a. m.

of the 13th, while at Burrwood the barometer was 29.72 at 3 p. m. of the 12th, and 29.71 at 7 a. m. of the 13th, a gradient of .01 of an inch in 100 miles. The oscillation of the barometer at Burrwood during the afternoon and night of the 12th indicated a tendency on the part of the cyclone to curve into Louisiana. Ocean going vessels in the eastern part of the Gulf of Mexico were so crippled by hurricane winds prior to the 12th that they could not send in their weather observations.

Observations at Galveston were of no material help. At 7 a. m. of the 13th the barometer was 29.79, and at 7 p. m., 29.68—a fall of .11 of an inch in twelve hours. The center of the cyclone passed through the Gulf, about 400 miles south of Galveston at 7 a. m. of the 14th, when the lowest barometer, 29.60 inches, occurred. The total fall in the barometer in twenty-four hours during the passage of the storm was .19 of an inch.

Barometer readings at Corpus Christi, the station nearest to the center of the cyclone when it moved inland, did not give any guidance until the death dealing storm tide was overrunning the coast region and leaving destruction in its wake. The barometer at Corpus Christi at 3 p. m. September 13th was 29.73 and at 3 a. m. of the 14th was 29.56, a fall of .17 of an inch in twelve hours. From this time until the cyclone moved inland, some distance south of Corpus Christi, the fall of the barometer was more rapid, and the barometer read 28.65 at 3 p. m. of the 14th when the center of the cyclone crossed the coast line. No deductions of a definite nature concerning the intensity and direction of movement of the cyclone could be made from the barometer changes.

Storm tides along the coast, however, carried a more distinct and decisive message, telling of the intensity and movement of the cyclone. The storm tide commenced rising at Burrwood during the night of the 10th-11th, was 0.3 of a foot at 8 a. m. and 1.1. feet at 8 p. m. of the 11th, at which time a storm tide of 0.3 of a foot made its appearance at Galveston. On September 12th 8 a. m. the storm tide was 1.7 feet at Burrwood and 0.7 of a foot at Galveston, Texas, a rise of 0.8 of a foot at Burrwood and 0.4 of a foot at Galveston in twelve hours. At 8 p. m. the storm tide was 1.9 feet at Burrwood, a rise of 0.2 of a foot in twelve hours, and at Galveston the storm tide had risen 0.9 of a foot in twelve hours to 1.6 feet. By the afternoon of the 12th the storm tide gave definite proof that the center of the cyclone was then moving towards the Texas coast west of Galveston. On September 13th, 8 a. m., the storm tide was 2.4 feet at Burrwood, a rise of 0.3 of a foot in twelve hours and was 2.6 feet at Galveston, a rise of 1.0 foot in twelve hours. The storm tide had commenced showing up at Aransas Pass (Corpus Christi) Texas, at this time, nearly thirty-six hours before the cyclone reached that part of the Texas coast. The storm tide had stopped rising at Burrwood, but at Galveston there was a storm tide of 3.6 feet, a rise of 1.0 foot in twelve hours, and a storm tide of 1.5 feet had showed up at Aransas Pass. The storm tides indicated that the center of the cyclone would move inland south of Corpus Christi, bringing a destructive and death dealing tide to the coast west of Galveston. We did not have authority to issue hurricane warnings, but we sent warnings to the Texas coast for "Gales and dangerous high tides," so worded to carry conviction and cause extraordinary precautions to be taken to protcet life and property. These warnings were sent form New Orleans to all Texas coast stations. We expected the forecaster in the Central Office at Washington to issue hurricane warnings for all stations from Galveston to Brownsville, but none was issued. Instead the following advisory message was sent from Washington to Galveston and Corpus Christi on September 13th:

11:30 p. m.: Disturbance apparently central in the Gulf south of Galveston. Barometer on coast steady with rising tendency last two hours, and is low over entire west Gulf. As center of disturbance can not be located watch barometer carefully during night and take all possible precautions against rising winds and high tides, especially if barometer begins to fall steadily."

(Sig.) Frankenfield.

The barometer at Corpus Christi was 29.62 when the advisory was issued, and at 7 a. m. was 29.37 inches, a total fall of .25 of an inch, less than .04 of an inch per hour which was not alarming. However, the storm tide was telling an alarming story, but no additional advices were received from Washington.

September 14th, at 3 a. m. the storm tide at Galveston was 7.6 feet, a rise of 4 feet in seven hours; at Aransas Pass (Corpus Christi), the storm tide was 4.0 feet, a rise of 2.5 feet in seven hours; at 8 a. m. Galveston had a storm tide of 8.7 feet, a rise of 1.1. feet in five hours, while at Aransas Pass the storm tide was 6.5 feet and rising at the rate of nearly half a foot an hour. These storm tides were causing destruction along the coast which continued until the center of the cyclone moved inland above Brownsville about 3 p. m. The storm tide banked up 12 to 16 feet in depth in the vicinity of Corpus Christi.

This ranks as one of the largest and most intense cyclones that has ever visited the West Gulf coast. However, due to the fact that the coastal area in the path of the cyclone was sparsely settled, and to the warnings, only 284 persons lost their lives. The property damage amounted to $20,000,000.00.*

---

* See Tropical Cyclones, Cline, The Macmillan Company, New York, pages 154-160, and 270-272, figures 6, 7, and 8. Also U. S. Monthly Weather Review September, 1919.

CHAPTER XXXIV.

TROPICAL CYCLONE, JUNE 21-22, 1921, TEXAS GULF COAST.

TROPICAL cyclones, notwithstanding they cover large areas, are sometimes lost in transit. A cyclone, first observed in the Caribbean Sea near Swan Island on June 16, 1921, followed an unusual course. It traveled across the Yucatan Peninsula into the Bay of Campeche, thence northward over the Gulf of Mexico, slowly curving towards the east and its center moved inland over Matagorda Bay, Texas, June 22, 1921.*

"Lost, Strayed or Dead" might have been applied to this cyclone for three or four days after it moved into the Bay of Campeche, June 18, 1921. The Weather Forecasters in Washington, who were supposed to keep track of tropical cyclones in that region, lost it when it moved into the Bay of Campeche, and discontinued issuing advisory warnings concerning its actions. However, the storm tide, on the afternoon of June 21st, gave the first warning we had that the cyclone was still very much alive and was approaching the Texas coast. Between 4 and 5 p. m. of the 21st, the official in charge of the Weather Bureau Office at Corpus Christi, sent a special observation to the forecast center, New Orleans, La., in which he reported a rapidly rising tide at Aransas Pass, near Corpus Christi. There was nothing in the observation except the tide to indicate that a cyclone was rapidly approaching the Texas coast.

The Corpus Christi barometer at the time of the special observation at 4 p. m., was 29.86 inches which was .02 of an inch higher than it was a 7 a. m. Based solely on the action of the tide, warnings were telegraphed immediately to all stations on the Texas coast, and the warning along with the

---

* See Tropical Cyclones, Cline, The Macmillan Company, New York, pages 161-165.

special observation on which it was based were telegraphed to the Central Office of the Weather Bureau, Washington. In this warning we stated that the cyclone was traveling towards the northeast. To my surprise the forecaster in Washington issued at 11:30 p. m. of the 21st, a warning contradicting in part the warning we had issued in the afternoon:

> Washington, June 21, 1921, 11:30 p. m.: Storm of unknown intensity central off mouth of Rio Grande moving northwest. Shifting gales tonight north of Colorado River, possibly as far east as Galveston. Every precaution should be taken. Advise all interests.

The cyclone did not travel towards the northwest, but it followed a course east of north, the direction in the warning we had issued.*

Conflicting advices in the warnings caused confusion and resulted in considerable complaint from citizens of Corpus Christi, Texas. Two forecast officials went from Washington to Corpus Christi to pour oil on the turbulent waters. These officials in their report to the Chief of the Weather Bureau deviated from the facts in the case, and stated that the blame rested on Cline because he had made an error in forecasting the direction of travel towards the northeast in the warning he had issued on the afternoon of June 21st, which had caused a misunderstanding of the warnings. This was in an effort to whitewash the forecaster in the Central Office whose warning issued several hours after mine had stated that the cyclone would move towards the northwest, whereas it did not move in that direction at all. Before making this report these officials should have checked the path followed by the cyclone as shown in the U. S. Monthly Weather Review. Here the path of the cyclone was shown to be towards the northeast as we had forecast in our warning. At this time the forecasters in the Central Office held to the view that the storm tide did not convey any information which could be used in forecasting the direction of movement and intensity of the cyclone.

---

* See report on Tropical Cyclone, June 21-22, 1921, by B. Bunnemeyer, Monthly Weather Review 49: 335. Washington.

Prof. Marvin, Chief of the Weather Bureau, evidently checked up on the movement of the cyclone and held up their report until I arrived in Washington six months later. Upon my arrival in Washington Prof. Marvin had their report handed to me and he personally told me not to deliver the report to the Chief of the Forecast Division but to hand it to him in person. Their misrepresentations were answered in a written report and I carried with me a copy of the U. S. Monthly Weather Review which showed the path followed by the cyclone was exactly as I had predicted based on the action of the storm tide. Prof. Marvin immediately summoned the forecasters into his office and told them in a firm manner that he wanted more accurate warnings for storms than they had issued in this instance. If the forecaster in Washington had weighed the information given by the storm tide he would have stated that the cyclone was moving towards the northeast.

This freak storm is discussed because it emphasizes the fact that a forecaster must be on the alert and utilize all information available that will aid him in giving the public the service to which it is entitled. The Washington forecasters responsible for the confusion created at Corpus Christi endeavored by misrepresentation to place the blams on me and failed. However, the official in charge at Corpus Christi, who had been alert and sent in the report on the storm tide on which the only advices worth while were issued was made the "goat", and forced out of the weather service.

CHAPTER XXXV.

STUDIES OF DESTRUCTIVE TIDES CAUSED BY TROPICAL CYCLONES.

METEOROLOGISTS, as far as I have been able to ascertain, had not at any time taken cognizance of the fact that the storm tides, and not the hurricane winds in the cyclone, are the death dealing forces that claim a toll of human lives and leave destruction in their wake. The fact that the winds in the cyclone cause the storm tide had been overlooked in the studies of tropical cyclones. The history of great tropical cyclones (hurricanes and typhoons) records enormous loss of life and great destruction of property by storm tdies, but it appears that prior to the Galveston disaster no studies of this feature had been attempted. Neither do we find where efforts have been made to give warning of the approach of the death dealing storm tides.

There are many records of disasters caused by storm tides such as that of the Calcutta cyclone of October 5, 1864. This was attended by a storm tide 16 feet deep over the delta of the Ganges and drowned 40,000 persons. The Backergunge cyclone of October 31, 1876, which caused an unprecedented storm tide, ranging in depth from 10 feet to nearly 50 feet over the eastern edge of the delta of the Ganges, drowned at the lowest estimates 100,000 persons. Along our own Gulf and South Atlantic coasts the loss of life from drowning by storm tides has been large where the cyclone moved inland over areas that were thickly populated.

During my spare time in the fifteen years following the Galveston disaster in 1900, I studied carefully the records about storm tides produced by tropical cyclones. These studies revealed that the destructive tides occur only in the right-hand front of the cyclone. The theoretical circulation of the winds in cyclones, as given in text books and other publications, could not cause the storm tides to occur in the manner in which they appeared on the coast in front of the cyclone. The only information found which dealt with the power of the winds in a tropical cyclone to produce swells was a diagram by Col. Reid, published in 1849, in which he showed the swells moving out uniformly in every direction from the center of the cyclone. This conformed to the accepted spiral inward movement of the winds around the cyclone center. Such a movement of the swells would not cause a tide along the coast of the magnitude which these cyclones produce. The action of the storm tides shows that Reid's representation is not correct. I was continually searching for facts about the wind directions in the cyclone which would enable me to solve this problem. No previous study was found that would help me chart the actual directions and velocities of the winds in the different parts of the cyclone so as to show their relation to the center and movement of the storm. This was essential in order to enable me to determine the physical forces in the cyclone that produce the storm tides.

Material for this study was available, but it had not been assembled and used for the study of this particular subject. Complete tide records from self recording tide gauges, made at stations along the Gulf and South Atlantic coasts, were furnished me by U. S. Army Engineers and the U. S. Coast and Geodetic Survey. The automatic tide records showed without question how the storm tides build up as tropical cyclones approach the coast. These showed that the storm tide builds up only in the right-hand front of the cyclone. What are the physical forces in the cyclone that cause the tides to build up in this manner? The accepted theories of the circulation of the winds around the cyclonic center, as shown in illustrations in text books and other publications, would not account for the storm tide building up only in the right-hand front of the cyclone. This problem had troubled me since 1900, and I was determined to find out what caused the tides to act as they did. The forecasters in the Weather Bureau, neither recognized nor studied the importance of storm tides, but devoted their efforts to forecasting hurricane winds resulting from the tropical cyclones.

The failure of the forecasters in the Central Office of the Weather Bureau at Washington to issue hurricane warnings for the tropical cyclone of September 14, 1919, the Corpus Christi hurricane, caused me to conclude that the progress I had made in this study should be published. I was convinced that it would prove of lasting benefit to humanity. I decided to submit the results of my work on this subject to the Chief of the Weather Bureau, and request his authority to publish it. At the same time I wanted his authority to continue my investigations on that subject with the further proviso that no one else in the service could take up the same line of research. I requested authority to prepare a study for publication on the subject, RELATION OF CHANGES IN STORM TIDES ON THE COAST OF THE GULF OF MEXICO TO THE CENTER AND MOVEMENT OF HURRICANES. More than a month passed and no reply to my request was received. About this time Prof. Marvin, Chief of the Weather Bureau, came through New Orleans on his way to Corpus Christi to make a personal survey of the effects of the hurricane of September 14, 1919. I asked him concerning my request for authority to study the storm tides, and he told me that the forecasters in the Central Office at Washington said there was nothing in my representations and that no relation could be established between the cyclone and the storm tide. Tide graphs and meteorographs had been prepared for two cyclones, the hurricane of September 28-30, 1915, and that of July 1-5, 1916. These showed the rise in the storm tide and the barometer changes as the cyclones approached the Gulf coast. The storm tides showed exactly where the center of the cyclone would move inland, while the barometer readings from all available stations would not give this definite information. Prof. Marvin examined the studies carefully and he authorized me to proceed with my investigations and publish my results. He remarked, "You certainly have here something of importance and value to hurricane forecasting."

Progressive rises in the storm tide as the cyclone approached the coast were easily shown on graphs. The main problem was to show the physical forces in the cyclone that caused the tides to act in the way they did. Cyclones do not travel in straight lines but their paths curve. Wind directions as related to the center and movement of the cyclone can not be represented on synoptic charts as had been the custom in former studies. The directions of the wind shown on the synoptic charts do not represent the actual wind directions as related to the center of the cyclone while it is traveling. The integration method had never been used in studying the directions of the winds in cyclones, and I introduced this method into the description of cyclones.

I brought out the fact that the winds in the right-hand rear quadrant of traveling cyclones have the same general direction as that in which the cyclone is traveling and continue so during the life of the cyclone. These are the winds that send out the swells that reach the coast always in the right front of the cyclone.

When the manuscript of my study, RELATION OF CHANGES IN STORM TIDES ON THE COAST OF THE GULF OF MEXICO TO THE CENTER AND MOVEMENT OF HURRICANES, was received by Dr. Charles F. Brooks, Editor of the U. S. Monthly Weather Review, he did not submit it to the Forecast Division, but proceeded to publish it in the U. S. Monthly Weather Review (March, 1920, 48: 127-146). I was told that when my studies appeared in print in the Monthly Review that the forecasters "hit the ceiling". They called Dr. Brooks to task and he replied, "You had already passed on the matter when you said there was no relationship. The study here shows that there is a definite relationship". They then said that the relationship shown in the study of the two cyclones in question was an accidental relationship and only a coincidence. This caused me to commence the collection and study of all the available material about tropical cyclones.

CHAPTER XXXVI.

MY BOOK "TROPICAL CYCLONES"

FORECASTERS and others in the Central Office of the Weather Bureau at Washington were positive in their assertions that my conclusions concerning the relations of storm tides to the center and movement of tropical cyclones were not scientifically sound. They freely expressed the opinion that the conclusions I had reached in the study of the two cyclones used were accidental and that similar conditions would not be found in the general run of tropical cyclones. I was convinced that my conclusions were scientifically sound, and was determined to prove beyond question that I had found something that would prove of great value to humanity. Weather observations and tide records in all tropical cyclones that had moved inland on the Gulf and South Atlantic coasts, from 1900 to 1924, inclusive, were assembled for further study and analysis to determine whether they would confirm my first findings. All meteorological data recorded in nineteen tropical cyclones were tabulated by use of the integration method, for each station over which the cyclones passed from the time the station entered the cyclone in front until it passed out in the rear of the cyclone. The wind directions and velocities and the precipitation were entered on charts showing their relations at each observation to the center and movement of the cyclone.*

In previous studies of cyclones haphazard methods have been used to describe the conditions which prevail in the different parts of the cyclone. Writers, when describing the characteristics of cyclones, used the points of the compass in designating a particular part of the cyclone area. It is well known that cyclones do not follow fixed directions, but travel towards

---

* For full details of this method see Tropical Cyclones, Cline, The Macmillan Company, New York, pages 26-34.

MY BOOK "TROPICAL CYCLONES" 183

every point of the compass. A cyclone traveling from the east towards the west would have entirely different conditions on its north and south sides from that which is found on the north and south sides of a cyclone traveling from the west towards the east. In fact the conditions would be reversed. In my studies a new and uniform nomenclature was adopted. This nomenclature divides the cyclone as follows: "front half", "rear half", "right half", and "left half"; then the quadrants of the cyclone are "right rear quarter", "right front quarter", "left front quarter", and "left rear quarter". This nomenclature can be used intelligently in any part of the world in describing conditions in different parts of the cyclone regardless of their direction of travel. The conditions found in the right rear quarter of a cyclone and in the other quarters are the same in all parts of the world regardless of the direction of travel. The quadrants of the cyclone are referred to, to the exclusion of the points of the compass, and this enables us to compare conditions in cyclones which occur in different parts of the world.

Certain officials of the Weather Bureau opposed the publication of my book without presenting any convincing criticism of the methods used and the conclusions reached. The complete text was finished and submitted to the Chief of the Weather Bureau in 1925, with request that it be published. Prof. Marvin, Chief of the Weather Bureau, referred the text to the scientific committee and requested them to show:

(1) Wherein the methods of analysis and presentation are not sound and legitimate.

(2) Wherein the observational data are not accurate, adequate, and definite.

(3) Wherein the inferences and interpretations of the evidence are not justified and correct.

(4) Where conflict arises with reference to preconceived ideas or current theories to show the advantage of such ideas or theories over the views, explaining the causes of the phenomena of the cyclone, advanced in this work.

Not a single one of the challenges made by Prof. Marvin was answered. These challenges were embodied in the Preface of the book and no one has yet answered them.

Authority was granted to me by the Secretary of Agriculture to publish the book privately. The book is well summarized in a review which appeared in the QUARTERLY JOURNAL of the ROYAL METEOROLOGICAL SOCIEY, (Great Britain) April, 1930, pages 203-205, as follows:

TROPICAL CYCLONES, by Isaac Monroe Cline, M.A., M.D., Ph.D., New York (The Macmillan Company), 1926. Pp. xxi, 301, Illus.

This book constitutes a notable advance in the collection and representation of precise data with regard to the tropical cyclones of the North Atlantic Ocean, observed at coastal and inland stations of the American continent. Tables are given showing the detailed hourly records and observations at from three to five stations in each cyclone during the time the stations were within the influence of the cyclone. A new method has been devised for charting meteorological data during the passage of a cyclone. The tables are given for sixteen cyclones discussed in the book, those which moved in on the Gulf and Atlantic coasts during the 25 years 1900 to 1924 inclusive. Of these one of the most interesting and valuable is the study of the cyclone of August, 1915, which entered the Texas coast on the 15th and passed over the region of the Great Lakes on the 21st and 22nd, having moved up the Mississippi Valley and traveled 2,000 miles from its coastal entry. Dr. Cline's work is, in fact, an exhaustive study along new lines of the distribution of wind direction and speeds, clouds and precipitation in the tropical cyclone. The author is particularly well qualified for the task he has set himself, having lived for 37 years, at Galveston, Texas, on the Gulf coast, and at New Orleans, in the capacity of meteorologist and district forecaster in the United States Weather Bureau. Under the delightful sub-heading of "personal Contacts with Tropical Cyclones", we learn that Dr. Cline has had actual experience of ten of these storms, which includes the most notable ones in the history of the region. The destructiveness of these terrible manifestations of nature needs no emphasis, but is brought home to the reader by the statement that a single cyclone has resulted in the loss of over 6,000 lives on the American continent.

Studies of cyclones have previously been based mainly upon synoptic charts which do not enable wind direction and velocity, the distribution of precipitation and cloud motions to be shown in relation to the cyclonic center. Even so recently as 1922 Sir Napier Shaw, writing in GEOPHYSICAL MEMOIRS No. 19 on "The Birth and Death of Cyclones" states that "we have no very satisfactory information about the distribution of precipitation in a tropical revolving storm". The method devised by Dr. Cline remedies such deficiencies. The path of the cyclone is first determined as carefully as possible by observation of the minimum pressures and shifts of the wind at stations both on the line of advance and on either side of it. A map being drawn of the track, a sheet of transparent paper is prepared representing the cyclonic area, divided into four quadrants by intersecting lines. On this the wind and cloud directions, wind velocity and precipitation are plotted hourly for from three to five stations; the center of the cyclone sheet being advanced each time along the track to correspond with the appropriate hourly movement of the storm. This method is approximately equivalent to greatly multiplying the number of observing points within the cyclonic area on a synoptic chart.

Following the discussion of individual cyclones are combination diagrams of wind, cloud and precipitation for a number of the larger and smaller storms taken separately, also a discussion of the cyclones which cease to advance and a general account, with thermograms, of temperature in cyclones. It is interesting to note that no polar front is indicated by these records. Observations of humidity are entirely absent throughout the book. Dr. Cline's "Summary and Conclusions" from the data afforded by the fourteen cyclones for which information is complete occupy pages 207 to 234 of his book, and it will not be possible to do more than outline them briefly here. The fundamental facts which emerge are the systematic differences in the inclination of the winds to the isobars and in the cloud movements in the four quadrants, and the localization of the precipitation in the front right-hand quadrant.

The winds in the rear right-hand quadrant blow in the same general direction as that in which the cyclone center is traveling and continue to do so during the life of the storm. These winds, with velocities of 40 to 100 m.p.h. or more converge upon the winds of the front right-hand quadrant, which have a variable inclination towards the center but which within 100 miles of the center blow nearly at right angles to the line of advance. The winds from the right-hand quadrant blow round the center and through the rear left-hand quadrant, whence they take a direction almost at right angles to the line of advance and finally curve somewhat sharply into and merge with the stream of the rear right-hand quadrant. Wind speed is much greater in the right half of the cyclonic area, and the greatest sustained speeds are found in the rear right-hand quadrant. The winds of the front right-hand quadrant have a somewhat special character blowing intermittently in sudden local and powerful gusts which appear to have a marked vertical component upwards. Clouds at all levels in the rear right-hand quadrant move in the direction in which the cyclone is advancing; in the front of the cyclone only the cirrus clouds do so; the lower clouds from which rain is falling, generally not more than half a mile above the earth's surface, are inclined more to the right than the surface winds. In the rear left-hand quadrant the lower clouds move mainly with the surface winds. Thus the larger cyclones travel in air currents which extend to the cirrus level. In cyclones less than 450 miles in diameter this is not the case, and the cirrus moves independently of the direction of advance. There is no evidence whatever of a spiral outflow of cirrus in all directions from the center in the upper air over the cyclonic area. It is pointed out that the average path of the prevailing winds of the region is a curve similar to that taken by the cyclones, and that a seasonal change of the one produces a corresponding change in the other. Neighboring high pressure areas affect both. Alto-cumulus and alto-stratas in the front part of cyclones move in a direction slightly outward from the center.

Precipitation occurs mainly in the front right-hand quadrant and almost ceases at any station with the arrival of the minimum pressure. In cyclones which cease to advance precipitation occurs both in the front and in the rear, but the area of lowest pressure shifts towards that of greatest precipitation, after which the storm soon dies out or loses the characteristics of a tropical cyclone. The source of moisture supply is through the winds in the rear right-hand quadrant, and the precipitation is due to the forced ascension produced by the convergence of these winds with those of the right-hand front quadrant. The position of neighboring land and water areas in relation to the paths of cyclones is discussed and can be shown to account in individual cases for the intensity of precipitation but not for its distribution, which is the same in all North Atlantic cyclones wherever they are observed.

The isobars are nearly circular from the center outward to 29.4 inches. Outside this line they show a spreading on one side and a crowding on the other, and this crowding and spreading shift their position as the cyclone changes its direction.

Dr. Cline therefore deduces that the primary cause of tropical cyclones is the convergence of air currents. He says "The air currents which converge and cause the air to ascend may be functioning in the upper air over the doldrums or on the border of the trade winds and may not always extend to the earth's surface." As a result of the rarefaction produced by ascension of air and precipitation the air flows in laterally and when at a sufficient distance from the equator the earth's rotation comes into force and gives birth to the cyclone, the convergence of the winds setting in last at the earth's surface. The cyclone persists by continuous redevelopment over the area of greatest intensity of precipitation, and the core of the cyclone is inclined from the area of lowest surface pressure forward into the cloud region over the area of maximum precipitation. As the cyclone travels the area of low pressure moves towards that of precipitation.

No such complete data as to the conditions obtaining in tropical cyclones have ever before been assembled, and Dr. Cline is to be congratulated upon the persistence and enthusiasm with which he has carried out his self appointed task. The book will prove of the greatest value to meteorologists in their search for the ultimate causes of cyclones, both of the tropical and temperate varieties. It should also be of great value as an aid in the forecasting of the destructiveness of a particular cyclone in particular localities. In this connection mention should also be made of the long paper by the author on "Relation of Changes in storm tides on the coast of the Gulf of Mexico to the Center and movement of Hurricanes", which is reprinted from the Monthly Weather Review of March 1920, as an appendix to the book. In this paper it is shown that the wind velocities and directions in different parts of the cyclone, as deduced from its ordinary symmetrical conception, will not explain the observed rises in the tides. On the other hand these rises agree well with the distribution of wind which has been deduced in the main part of the book. The largest and longest waves are created in the rear right-hand quadrant; these waves pass on through the cyclonic area and move to shore, where they cause a rise in the water in front of, and to the right of, the line along which the cyclonic center is advancing. The rise in the water begins when the center is 300 to 500 miles distant and continues till the cyclone crosses the coast. The rises occurring on the coast near the center of the storm are from 8 to 15 feet above mean sea level. This deduction should also have practical value.

Valuable features of the book are a number of quotations from other authors who have written on the subject, and the bibliographies to the main part of the book and to the appended paper. The book is well got up, and the numerous diagrams are clear. No errors or misprints have forced themselves upon the notice of the reviewer, with one exception: on page 235 under item (3) of the bibliography "Geographical" should read "Geophysical."

(Signed) E. W. Barlow

In looking up the biography of Dr. Barlow we find that he is "Fellow Royal Meteorological Society", "Fellow Royal Astronomical Society", and an official of the Meteorological Office, Air Ministry of Great Britain.

The question is frequently asked. Can the atomic bomb be used successfully to break up and dissipate a tropical cyclone? The answer is no. A tropical cyclone is always of considerable extent. It is developed by the physical forces of nature acting under favorable conditions and the energy expended in keeping it in operation is so great that it can not be estimated.

CHAPTER XXXVII.

WIRELESS TELEGRAPHY—RADIO

HAVING seen the great advances made in the development of the telegraph and telephone, I was much interested in any promising invention that would add to or improve communication. Investigations which had for their object the transmission of messages without wires were being conducted along with the development of the telegraph and telephone. William Marconi, of Bologna, Italy, began in 1890, at the age of sixteen years, experiments to test the theory that the electric current is capable of passing through any substance, and if started in any given direction, of following an undeviating course without need for a wire or other conductor. Marconi came to the United States in 1899, and displayed his method of wireless in 1900. In 1902 he sent wireless telegraph messages across the Atlantic Ocean. He then organized a subsidiary of the European Marconi Wireless Telegraph Company in the United States.

Inventors in the United States had been working on this idea, too, and patents for wireless telegraphy had been taken out in this country, based on the same principles employed by Marconi. The De Forest Wireless Telegraph Company was organized in the early 1900's. The Signal Corps, U. S. Army, was giving special attention to the development of wireless telegraphy. About 1904, Col. William A. Glassford, Signal Corps, U.S.A., who had been one of my instructors at Fort Myer in 1882, visited me in New Orleans. He was then under instructions to go to Alaska to set up for the United States Government, wireless telegraph stations in that region. He predicted then that the growth of wireless transmission would be phenomenal and that in time wireless would supersede the telegraph and telephone. I had the utmost confidence in his judgment. I immediately invested $500.00 in stock in the De Forest Wireless Telegraph company. The United Wireless Telegraph Company was organized later and absorbed the De Forest Company. De Forest stockholders exchanged their stock for United Wireless Telegraph Company stock. The growth of the United Wireless Telegraph Company was rapid. Subsidiary companies were chartered in different States of the Union. The outlook was most promising. The United Wireless outdistanced the Marconi company which failed to get a good foothold in this country although progressing rapidly in Europe.

About the year 1912, the Marconi people planned to wreck the United Wireless Telegraph Company in one day by executing a coup. The Marconi company suspected that officers of the United Wireless Company were selling their personal stock in the company through the mails, representing it as promotion stock. This if proved would show that they were using the United States mails to defraud the public. They planted a stool pigeon in the offices of the United Wireless Telegraph Company for the purpose of collecting evidence which could be presented to the United States Postal authorities. Evidence so strong that it could not be questioned was secured. The Marconi people took precautions against a leak, and set a specific date for the United States Postal authorities to raid United Wireless headquarters. In advance of this date the Marconi people had employed and instructed lawyers in every state in the United States where the United Wireless Company was incorporated to apply immediately for receiverships in those states when they received telegraphic notification that the United Wireless Officers had been arrested. On the "day of judgment" for United, receiverships "took over" the same day in several states before the stockholders of the United Wireless Telegraph Company knew what was happening. Under the receiverships the United Wireless Telegraph Company was soon bankrupted.

It seems to me that in cases of this kind there should be some law enforced in the United States Courts which would require a delay in the appointment of receivers until the stockholders in a company could be given due notice, so that they could have some say in the matter. From my observations it also seems that stock companies have been thrown into receiverships with the planned purpose of bankrupting them so that assets could be bought at a ridiculously low price. This was apparently the object of the Marconi Company in this instance.

Marconi Wireless representatives succeeded in having the assets of the United Wireless Company appraised at such a small value that it was ridiculous. They had planned to get control of United Wireless without reckoning that the United Wireless stockholders would be in on the "kill". By paying off the appraised value of the patents, plant, and stations, the stockholders of the United Wireless Company had prior right to take over the entire assets, and that is what they did. Some of the large stockholders in the United Wireless Company arranged through an attorney to collect a small amount on each share of stock. A sufficient sum was collected to pay off the appraised value of the United Wireless Telegraph Company; the attorney with authority from the Stockholders of the United Wireless Company took the money, presented it with his power of attorney to the United States Court, and the assets were turned over to the attorney for the United Wireless stockholders. The Marconi people recognized that they had been outwitted, and they proposed to the United Wireless stockholders that the two companies unite under the name of the American Marconi Wireless Telegraph Company, the stockholders in each company to get stock in the new company of equal value to that previously owned. This was done and all litigation was stopped.

During World War I both General Electric and Westinghouse Electric developed and patented some wonderful improvements in wireless telegraphy. When the war ended, European officials of the parent Marconi Wireless Telegraph Company sent representatives to the United States with authority to buy these important patents. Uncle Sam stepped in and said, "No!" The United States Government told General Electric and Westinghouse to proceed to organize a company to take over all wireless and radio patents with requirements that 80% of the stock be owned by citizens of the United States. The Radio Corporation of America was organized in 1919 for this purpose. Stockholders in American Marconi exchanged their stock into

Radio Corporation of America when it took over the American Marconi Wireless Telegraph Company. Radio Corporation of America made rapid progress; in fact it developed more rapidly than did the American Bell Telephone. In 1924, the common stock was advancing in value rapidly. I sold my Government bonds and bought 600 shares of the common stock at $30.00 a share, with the intention of holding it as an investment. The stock advanced rapidly. It seemed that my wealth increased by hundreds every day. In January 1925, Radio was $75.00 a share, $45.00 a share more than I had paid. It then dropped back to $65.00 a share. I decided that the stock was not worth more than that amount and sold out at that price. I made a good profit and the money was put where it helped others. The stock then started up again and went by leaps and bounds until it reached $525.00 a share. If I had carried out my original intention I could have cleaned up nearly three hundred thousand dollars in four years. However I am satisfied with having sold. The stock was later split one share into twelve while it was at the top. I now own Radio stock for which I paid $3.00 a share which is just a little above the price I paid for the old stock in 1924. The stock is now selling around $10.00 a share which is equivalent for $120.00 a share of the old stock which I sold for $65.00 a share 18 years ago.

Wonderful progress is being made in electronics, and we can not imagine what the future will bring. Television is coming soon, and as time goes on the possibilities it offers seems unlimited. The telephone and telegraph have made wonderful strides, but they are being far outstripped by wireless devices that use the air waves.

Flood Crest Transferred from River Channel to Overland Flood.

CHAPTER XXXVIII.

AMERICAN PORTRAITS—SECOND AND THIRD COLLECTIONS.

FOLLOWING the sale of my first collection of American portraits in January, 1918, I continued to devote my recreation time to buying and restoring old paintings, the majority of which were portraits. In 1925 I had a collection of American portraits, mostly by American painters. This collection was sold to a New Orleans collector for $18,000.00. There were paintings by such artists as Gilbert Stuart, Thomas Sully, John Wesley Jarvis, Matthew Harris Jouett, Samuel L. Waldo, S. F. B. Morse, G. P. A. Healy, J. G. L. Amans, A. Rinck, and many other outstanding American painters along with some European painters. The purchaser took out some old Masters that were in the collection for himself, and sold the others at public auction. The sale was under the management of Samuel T. Freeman & Co., Philadelphia, Pennsylvania. While I was not interested financially in the paintings in any way, the illustrated catalogue of the sale bore the title "The Dr. I. M. Cline collection, of New Orleans, Louisiana." Mr. Mantle Fielding wrote the foreword in which he said:

> "The pictures have not been culled from various sources as is often the case, but have been the result of the efforts of one man, Dr. Cline of New Orleans, Louisiana, who through years of study, and with a very real love for the beautiful, found delight in seeking worthy examples of early American art in the aristocratic old homes of the South, and more particularly among some of the most exclusive families of Louisiana."

Mr. Charles X. Harris, who represented the purchaser of my previous collection, was present at the sale and purchased some of the portraits. He got the impression, from the heading of the catalogue, that the paintings were my property and were being sold for my account. He entered, in the illustrated catalogue, the price that each painting brought and sent it to me for my information. This catalogue is highly prized because it shows the value placed on American portraits by the general public bidding at auction. A portrait of Abraham Touro by Gilbert Stuart, for which I paid $325.00 at auction on Royal street in New Orleans, Louisiana, brought $5,000.00 at the auction sale in Philadelphia, Pennsylvania. Some of the paintings which the collector did not sell at auction were later given to the Isaac Delgado Museum of Art, New Orleans, Louisiana.

Another collection of American portraits, the third which I had brought together and restored in my recreation time, was, sold to the **Montgomery Museum of Fine Arts**, Montgomery, Alabama, at a price which covered original purchase cost and restoration. In addition to the paintings sold to the Montgomery Museum of Fine Arts, I gave them some paintings. Among them was an unusual portrait of Judah P. Benjamin at the age of about 35 years, and also "Harmony in Black and White" by Luis Graner, which shows Graner and his baby boy. It was painted while Graner's wife was in a hospital and not expected to live, hence the sadness on the face of Graner while the baby sleeps serenely.

Dr. Robert Glenk organized the Louisiana State Museum, using as a nucleus the material which came back from the Louisiana exhibit at the Saint Louis Exposition in 1904. Dr. Glenk established a wonderful historical museum without any great expense to the State of Louisiana. I had the pleasure of cooperating with Dr. Glenk in bringing together a collection of portraits representative of American artists covering the early period of the development of American art. It is probably the most representative collection of portraits by American painters to be found in the United States. Among portraits which bear my name as having given them to the Louisiana State Museum, is the life-sized portrait of John Slidell by E. Wood Perry, Jr., N.A., painted while Slidell was United States Senator from Louisiana, also the portrait of Madame Pauline Boyer, life-size standing by her harp, painted by de Havilier about 1820. Madame Boyer was a noted harpist

AMERICAN PORTRAITS—2nd AND 3rd COLLECTIONS 195

and had entertained the crowned heads of Europe with her music before she came to New Orleans where she soon became the outstanding leader in musical circles.

Portraits of Lucien Bodro, painted by A. Rinck, 1856, and of Pierre Landrieux by J. G. L. Amans, 1845, were given to the Isaac Delgado Museum of Art, New Orleans, Louisiana. Bodro entertained Jenny Lind and Charles Dickens when they visited New Orleans. A large mountain landscape "Morning", by W. Keith, was also presented to the museum.

There was romance in looking up the history of artists as well as of the persons who were represented by the portraits. One of the most interesting was in connection with a portrait of Robert Nicolas Charles Bochsa, painted by Peter Capman. Bochsa was born in 1789. He was harpist to Emperor Napoleon at the age of 13 years and continued in that capacity for a few years. He was also harpist to Louis XVIII in 1816. He had already distinguished himself as a musical composer. In 1817 he became involved financially and fled from France. He was tried **in absentia,** found guilty and sentenced to twelve years imprisonment. Bochsa took refuge in London, where he was lionized, especially by the ladies. He gave concerts in every court in Europe, except that he never dared go back to France where prison bars always awaited him. In the early eighteen thirties he eloped from London with the wife of Sir Henry Bishop. They stopped in Louisiana where a portrait of Bochsa at his harp was painted in 1837. From Louisiana Bochsa went to Australia where he died of gout in 1856.

Peter Capman, the painter, was born and studied art in Copenhagen, Denmark. He painted distinguished persons at European courts, and in this way became acquainted with Bochsa who entertained at these same courts with his music. In the early 1830's Capman visited America where he renewed his acquaintance with Bochsa, and painted his portrait. Capman then went to the Island of Guadaloupe in West Indies to execute portrait orders there. There he met a Spanish Princess who had fled from Spain to Guadaloupe to escape being placed in a convent because of her refusal to marry the Prince whom her family had selected for her. A love affair developed between Capman and the Princess. They could not marry because Capman was not a nobleman. Friends of Capman who knew of the affair took the matter up with the Danish Government and succeeded in having Capman created a nobleman, so that he could marry the Princess. They were married, the Spanish Government relented, and they were invited to the court of Spain. They boarded a ship in the harbor of Guadaloupe to return to Spain, but before the ship was ready to sail, an earthquake occurred and all on board, including the Prince and Princess Capman, were drowned.

Note:—For the information of those who are interested in artists who worked in New Orleans reference is made to— ART AND ARTISTS IN NEW ORLEANS DURING THE LAST CENTURY, 1920, and also, CONTEMPORARY ART AND ARTISTS IN NEW ORLEANS, 1924, both by Isaac Monroe Cline. They were printed in pamphlet form and also appeared in the Annual Reports of the Louisiana State Museum for those years.

CHAPTER XXXIX.

THE GREAT MISSISSIPPI RIVER FLOOD OF 1927.

CONDITIONS over the Mississippi Drainage Basin during the Winter and Spring of 1927, indicated the approach of high water in the lower Mississippi Valley. Careful watch was kept on all weather conditions which might favor the building up of a flood of unusual proportions. Mr. R. A. Dyke, assistant forecaster on duty, issued forecasts early in April for a flood of dangerous proportions. The warnings and advices were given wide distribution daily by maps and bulletins, and complete information was furnished to the newspapers. However, the newspapers were publishing these flood warnings in obscure places where the people would not see them unless looked for. Furthermore, important flood information which we furnished the press for the benefit of the public was not being published.

Flood conditions in the central Mississippi Valley had assumed such serious proportions by the middle of April, that a dangerous flood was certain to occur in the lower Mississippi Valley and great destruction and loss of life would result unless precautions were taken to protect life and property. Action on the part of the people to meet the flood and combat it could prevent at least part of the threatened damage and prevent loss of life. On April 14th, newspaper reporters who were handling the flood warnings and advices of the Weather Bureau were called into my office for a conference. I asked them why the newspapers were not publishing in full flood warnings and other important information which the Weather Bureau was issuing. They informed me that the merchants of New Orleans had a censorship committee handling flood matters. Among the duties of this committee was that of the suppression of the publication of flood news because such flood news would prevent country merchants from coming to New Orleans to buy goods. "Billy" Phaff was chairman of that committee. He was a close personal friend of mine and one of nature's noblemen. I got Mr. Pfaff on the telephone at once and said to him:

> "The greatest flood in history is approaching and your action in suppressing flood information and warnings which we are giving the newspapers to help the people get ready for his flood is jeopardizing the lives of men, women, and children. Your committee is playing with fire and a flame of public indignation will flare up against you should your action, in contributing to the possible loss of life and property, become public. I am telling you now the people of the lower Mississippi Valley are going to get the warnings from the Weather Bureau. You may control the press, but we have the mails, the telegraph, the telephone, and the radio, and you cannot suppress the distribution of flood warnings and information through these channels. We are going to see to it that the people living behind the levees are warned that they and their property are threatened with great danger."

Pfaff reported to his committee my conversation with him. The Association of Commerce took the matter in hand. They telephoned me that afternoon, requesting me to meet with them the following day at 10 a. m. I told them that I would meet them at the hour named.

Flood forecasts and warnings issued by the Weather Bureau had been so successfully censored up to this time that few people in New Orleans realized that a flood of great proportions was almost upon them. The people had been in the habit of relying on the newspapers for weather bureau forecasts and warnings and the censorship had deprived them of this information. The committee stated that it had not been their intention to suppress publication of the weather bureau forecasts and warnings of the flood but to prevent the publication of general flood news which might be misleading. Their intentions were good but their actions had carried too far, and they soon learned that they had been playing with fire that spread to a conflagration. The newspapers immediately gave prominence to the flood warnings of the Weather Bureau and the sudden burst of information concerning the flood caused a panic.

THE GREAT MISSISSIPPI RIVER FLOOD OF 1927   199

Organizations which have for their object the rendering of aid to the public in case of calamities should never take action when weather conditions are involved without consulting Weather Bureau officials. Without consulting me, the flood forecaster, the relief organizations proceeded to establish refugee bases in the city of New Orleans. The refugee bases were supplied with skiffs which were to go through the imaginary flood and bring in the flood sufferers. Blankets and safety pins were provided so that those who were wet could be wrapped up until dry clothing could be provided. There was no necessity for such action, which was taken without the knowledge of the Weather Bureau forecasters. Had we known that such action was being contemplated we would have told them that it was not necessary, for we knew that New Orleans would not be flooded. The flood situation was such that I knew that the levees could not carry the flood waters as far as New Orleans. I knew that the Bayou des Glaises levees could not be raised high enough to prevent the flood waters from running over their tops and washing them out. However, I could not say this publicly because it would have been construed as a criticism of the levee engineers. Many people filled their attics with food supplies to carry them through a flood that was never forecast and would not occur. I reasoned with many people against such action but the mass psychology of fear could not be checked but spread rapidly. Thousands of people left the City daily. I was told that many of them were drawing their money out of the banks.

Bankers and merchants organized and asked that the levee at Poydras, below New Orleans, be dynamited to form a spillway and prevent New Orleans from being overflowed. Approval of the authorities in Washington to dynamite the levees was obtained. However, the people who resided in and owned property in the area which would be overflowed by the Poydras spillway prevailed on Governor Simpson to withhold his approval for the opening of the Poydras spillway. Simpson told the bankers and merchants of New Orleans that he did not consider that it was necessary to open the levee at Poydras. Lem Pool, chairman of the banker's committee, called me on the telephone and plead with me to go to Governor Simpson and tell him that the levee should be cut at Poydras in order to save the City of New Orleans. I told Mr. Pool that I did not consider New Orleans in danger from overflow and I could not go to Governor Simpson with such a request. Pool then said: "Dr. Cline, the people of New Orleans are in such a panic that all who can do so are leaving the City and it is ruining business." I told him that I was sorry I could not do something but I did not see my way clear to go to the Governor. I returned to my desk with Pool's appeal ringing in my ears I knew the people were leaving New Orleans in large numbers every day and I was anxious to help allay the unnecessary fears of the people. The opening of the levee appeared to be the only solution that would restore confidence. I called Pool on the telephone and said: "You may go to Governor Simpson and tell him that there is another rise in the river on its way here and that if the levee is going to be opened to relieve the situation it should be opened at once." The following day trucks were quietly sent into the area which would be overflowed by the spillway and the people and their belongings were moved into New Orleans. My duty was to forecast the flood and advise the people of danger; the levees were under another branch of the Government service and I could not say what the flood would do to the levees. I could only say "If the levees hold the volume of water now in sight".

Mr. R. A. Dyke, assistant forecaster, was on forecast duty during April, and did an excellent piece of forecasting channel floods as the water reached higher stages.

Levees had held up well through April. No crevasse had occurred which would influence the flood in the New Orleans district. However, I knew that the levees could not carry the tremendous volume of flood waters then in sight within the river chanels between Louisiana and Mississippi, and that there would be crevasses which would cause disastrous overland floods. The Glasscock levee, Concordia Parish, Louisiana washed out Saturday afternoon, April 30. This water would unite with the flood in the channel at Old River, the overland flood joining the flood in the Mississippi proper, and this would give a much greater flood than had previously beeen forecast for the New Orleans district. It was evident that the levees along Bayou des Glaises could not be raised high enough to prevent the flood waters from overflowing them and washing them out, and this would be disastrous to the inhabitants of the valleys above New Orleans, particularly to the west of the Atchafalaya. On Sunday, May 1st, I issued the following extraordinary flood warning:

> The water flowing through Glasscock crevasse will return to the main stream, joining the floods from the Ouachita and Red rivers, and intensify the flood situation in the Atchafalaya and Mississippi rivers, below Old River. Every precaution should be taken against the following stages: Angola, 58.5 to 59.5; Baton Rouge, 49 to 49.5; and on the Atchafalaya, Melville, 48.5 to 49.5, May 10th to 18th.

These stages were 1.5 to 2.5 feet above those given in the forecast issued the previous day, April 30th. This placed the crest of the flood 4.2 feet higher than the previous flood record of the Bayou des Glaises region. The predicted height of the flood was considerably above the tops of the levees along Old River and the Bayou des Glaises. While I knew the levees could not be raised to meet this flood, I could not say that such was the case, because that would have been taken as a criticism of the levee engineers. The flood warning was telegraphed immediately to Hon. John M. Parker, National and State Flood Coordinator for Louisiana, at Baton Rouge. This extraordinary flood warning was sent to every telephone in the populous area behind the Bayou des Glaises and Old River levees with the instructions to give the information to every person residing in that region. Mr. Parker called me on the telephone and asked me if the stages telegraphed him were correct. I told him that the stages he repeated to me were the correct forecast. He then asked me if I was certain those stages would be reached. I told him that if the levees could be made to carry the water in sight that the stages forecast would be reached. Mr. Parker replied, "Levees can not be raised by building on top during the next two or three weeks to carry those stages." I informed him that I had sent the warning to every telephone in the region below the Old River and Bayou des Glaises levees, with the request that all persons be advised concerning the seriousness of the situation. He then said: "I will telephone the authorities in that area and tell them what you say and urge them to move the people and their belongings to places of safety as soon as possible." The warning had the desired effect. May 14th, just two weeks after the warning was issued, several miles of the Bayou des Glaises levees were overflowed and suddenly washed away, turning loose a deluge of water like that from a bursted dam. This wall of water, 8 to 14 feet in depth, rolled down over a region which two weeks previously was thickly populated. Buildings were swept from their foundations but the people had all been moved to places of safety, and, although it was the greatest onrush of water in the history of this section of the country, not a human being was drowned. The warning saved thousands of human lives. Previous floods from smaller crevasses had drowned hundreds of people because they had not received warning in time to permit of their rescue.

I had studied the characteristics of overland floods, but I had not reduced my conclusions to writing. This knowledge of overland floods was my guide in the issuance of the extraordinary flood warnings of May 1st. I had found that overland floods travel at about one third the rate of travel of channel floods. I had made it part of my work to check every feature of the movement of flood waters so that I could give reliable advices to the public I was serving. I had learned that "knowledge is power", and that success can only be achieved by acquiring all the knowledge possible relating to the various features of the work in which I was engaged.

Soon after I went on forecast duty in May, 1927, the newspapers carried a statement that river engineers were of the opinion that the flood crest, then at Vicksburg, would with the ordinary rate of travel of flood crests—40 miles a day—pass New Orleans within the next ten or twelve days. Flood crests when in the channel of the river do travel at the rate of about 40 miles a day. This flood crest was no longer in the channel of the Mississippi River, but crevasses had transferred the flood crest out into the Tensas Basin. We now had to deal with an overland flood, the rate of travel of which is quite different from channel floods. I realized that Hon. Herbert Hoover, who was the representative of President Coolidge, in the flood area, would need definite information in this respect for guidance in his flood and rescue work. On May 10th, 1927, I prepared and published a special bulletin dealing with overland floods and their rate of travel. This bulletin showed that overland floods travel 14 to 15 miles a day, river distances; this was making history. Overland flood crests travel down stream only one third as rapidly as the crest of the flood travels when in the channel of the river. The crest of this flood now being opposite Vicksburg out over the Tensas Basin had to be considered as an overland flood independently of the channel flood. Warnings and advices had to be based on the movement of overland floods and not channel floods.

Immediately on completion, the bulletin giving the rate of travel of overland floods as related to channel floods was dispatched to Secretary Hoover at Baton Rouge. A long distance telephone call came back from one of Mr. Hoover's secretaries. He said, "Mr. Hoover desires a conference with you, and asks me to say that he regrets that he will not have time to visit you in your office and requests that you meet him in his car at the Union Station at 8:30 p. m. when he goes through New Orleans." This was the opportunity I wanted, for I felt that I could render great service to the people of this area through him. I was at the station to meet him when the train came in. Several important persons who were acquainted with my success in forecasting disasters had told him of my previous flood work. Secretary Hoover took me into his private office, and we went over the flood situation. I pointed out to him what the flood would do in the next two weeks and discussed the forecasts with him in their connection with future relief work. He called me into conference with him each time he went through New Orleans and had my views concerning flood movements and crest stages charted on maps for ready reference.

When the Bayou des Glaises levee washed out May 14th, the following special bulletin was broadcast by radio, telephone, telegraph, and mail at 5 p. m.:

> The head (depth) of the rolling flood of water is 16 feet at Kleinwood and 6 to 8 feet in depth at other places. The outflow from these crevasses will pass down west of the Atchafalaya River and inundate the greater portion of the lower Avoyelles Parish, Saint Landry east of the Teche, Saint Martin, eastern Iberia to the Teche, and may overflow the lower Teche with indications pointing to one of the greatest floods in the history of that region including Morgan City.

Following the issuance of this bulletin the Red Cross called me on long distance telephone, and asked my advice about locating a refugee camp at Saint Martinville. I told them that they had better open the refugee camp at Lafayette. The inhabitants of Saint Martinville rose up in arms and declared they would not have flood waters, for they had never been flooded. Their local newspaper came out with an editorial criticising my action as unjust to Saint Martinville, and assuring the people that Saint Martinville would not be flooded.

There were no gauge relationships for use as a guide in the preparation of warnings and advices to the people telling them how deep the water would be and what to do to save life and property. But some 300,000 people were threatened with the most disastrous flood in modern times, that is, in the United States. The greater number of these people would have to be told to move out with their belongings to places of safety. However, there were many who could be assured that they could remain in their homes and not suffer from the flood. We had nothing to guide us in preparing warnings and advices except the deluge of waters rolling down the valley. We had to work out new premises and build a forecast structure along new lines. Flood forecasts would have to be based on new methods. Profile maps of the region subject to the overflow were absolutely necessary. The State Board of Engineers told us they had no profile maps of that region. We then applied

THE GREAT MISSISSIPPI RIVER FLOOD OF 1927

to the United States Engineers, and they had no profile maps of the region threatened with inundation.

Profile maps had to be secured in some way. The elevations of the ground, above sea level, for the Atchafalaya Basin were absolutely necessary for our use, so that we could warn the people in the threatened area what depth the water would reach above the level of the ground. The ground was undulating. In some places the onrush of water would be so deep that homes would be washed away, and in many places, where homes would not be washed away, the depth of the flood would gradually undermine houses and damage them. All people residing in such areas must be told to move out. Adjoining localities, which to the ordinary observer appeared threatened, would not be flooded and people could remain in their homes without suffering material damage. These different localities had to be located and charted so that we could warn the people correctly.

The interests of 300,000 people, residing in one of the richest sections of the United States, would be affected in some way by flood waters, and this called for a service that had never before been attempted by anyone anywhere. The date on which the flood would arrive and the depth the water would reach over the ground was absolutely necessary to be known before warnings and advices could be given that would enable rescue workers to move inhabitants and their property from areas where the flood would be destructive to life and property. On the other hand people residing in areas where the depth of the water would not be dangerous must be told to remain in their homes where they would be safe, as the flood would not reach them.

This was the situation confronting us preceding the approaching calamity. We hit on an idea that enabled us to make topographic maps. Officials of railroads, which had trackage in the area which would be flooded, were asked for profile maps of their railroad beds and tracks. They acted promptly. They telegraphed their headquarters to rush these maps to New Orleans, and the maps were immediately turned over to us. This gave us bases from which we could work out the elevations of the ground above sea level in many localities. We could then construct rough contour maps which would enable us to predict the depth the water would reach above the ground in given localities.

We did not stop for sleep, but worked day and night until the profile maps were ready for use. Delay meant probable loss of human lives and property, and this we were determined to prevent. Credit for working out the ground elevations and making rough contour maps from railroad elevations belongs exclusively to the officials of the Weather Bureau at New Orleans. None of the engineers, State or National, assisted us in this work. Mr. R. A. Dyke, assistant forecaster, and W. F. McDonald ,First Assistant, did this work under my supervision. We then furnished the information to engineers and other agencies, working in the flooded area, for their use in carrying out instructions contained in our flood warnings and advices. In this way some persons got the impression that the engineers worked out the profile maps, but they did not have any part in it.

Overland floods had been studied previously and their rate of travel through different surface conditions determined. We now had the rate of travel of overland floods, the elevations of the ground, and the height of the flood crest then in the Tensas Basin above sea level, and were ready to issue definite warnings and advices. We determined the volume of the flood waters, and this, with the foregoing information at our command, we were able, for the first time in history, to issue warnings giving the rate of travel of an overland flood, and the varying depths which the water would reach as related to the ground. This was definite information which the inhabitants of the valleys could use to advantage.

Decisive action was necessary in so serious a situation. I assumed the role of dictator in giving, not only the advices as to the height of the flood, but in telling people what they must do in order to protect life and property. I told the people what they must do and what they must not do. Warnings and advices were issued in terms that carried understanding and conviction. Some of the warnings given in that memorable flood follow:

May 18th:—Special bulletin—We consider it advisable to evacuate (move everybody and everything out of) that area 15 to 20 miles wide west of the Atchafalaya to the Baton Rouge - Lafayette Branch of the Southern Pacific Railroad. Conditions also look serious for all towns on the Teche from port Barre to Breaux Bridge, and be ready for evacuation if the situation demands.

Note—The flood reached Breaux Bridge May 26th with the water five feet deep and much greater depths prevailed in other places.

May 19:—Flood water should reach 8 feet above the rails of the Southern Pacific Branch line between the Lafayette highlands and the Atchafalaya River, and overflow the low lands behind Bayou Teche below that line down to and around Saint Martinville, with water probably entering that town May 25th to 30th.

Note—Floodwater entered Saint Martinville on the morning of May 25th, and on the 26th the town was flooded with the water two and a half feet deep over the railroad track at Port Street.

May 20th:—The Atchafalaya bottoms of Saint Martin and Iberia Parishes should be evacuated as soon as possible . . . Back water will overspread Iberville and other parishes between the lower Atchafalaya and Bayou LaFourche, but the rise in the water will be slow so that the evacuation of the territory being inundated by back water in that region is not considered advisable until the depth of the water in individual localities makes it necessary to move people. No credence should be placed in rumors that the Mississippi River is changing its course into the Atchafalaya Basin at Old River.

May 21:—Water will probably come down west of Bayou Teche and cause flood stages in the Vermilion River near Lafayette. . . . The Teche will have more water than it can carry and will probably overflow its banks in the vicinity of New Iberia by June 1st.

May 22nd:—The Teche will probably overflow its banks at New Iberia, although no serious flood conditions are indicated in that locality.

New Iberia was later told that the flood water would reach the Southern Pacific track near the depot. The people of New Iberia were so certain that the flood water would not reach that locality that they employed a private flood expert to give them a statement which then could be given to the press, that the flood waters would not reach New Iberia; this statement was printed in many newspapers.

While I was at home having dinner about 6 p. m. on May 22nd, a long distance telephone call came in from Jeanerette, Louisiana. The telephone operator at Jeanerette told me that she was speaking at the request of Senator Provost who was so excited he could not talk over the telephone. It was Sunday and the flood rescue workers had gone into Jeanerette without proper authority and placarded the homes, telling the people that the rescue trucks would be in Jeanerette Monday to move them and their belongings to the rescue camp at Lafayette. The rescue workers showed no sense of judgment, as they had not conferred with me. I told the operator to say to Mr. Provost that Jeanerette would not be flooded and the people could remain in their homes with safety; that I would go to the Weather Bureau Office at once and send him a telegram confirming what I had told her, and in it would tell the people of Jeanerette to stay in their homes and refuse to move. The people of that town refused to move, stayed in their homes and had no flood water.

Another bulletin:

May 23rd:—No action should be taken to evacuate towns on the main line of the Southern Pacific Railroad.

During the forenoon of May 24th several persons from Lafourche and Terrebonne Parishes called me on the long distance telephone, and plead with me to say that the flood situation in that region would be relieved by dynamiting the embankment of the Southern Pacific Railroad. I told them that their viewpoint was wrong and advised them against such action. The pleas were so urgent that a special bulletin was issued at 1 p. m.:

> May 24th:—It has been suggested that the main line of the Southern Pacific Railroad be opened to help relieve this flood now entering the Atchafalaya Basin. We can not see where any advantage can be gained by such action as it appears that the continuous operation of this railroad is in the best interests of the general public and of the residents of the flooded area, offering benefits by continued operation out of all proportion to the comparatively slight measure of relief that could be afforded by the proposed action.

Mr. Watkins, Vice-President of the Southern Pacific Railroad, later told me that explosives had been placed along the roadbed, and that, had it not been for this bulletin, the roadbed would have been dynamited that afternoon. The bulletin was sent at once to that locality, and Mr. Watkins said that on reading it the men picked up the dynamite, departed and no further threats were made. Railroad officials had been pleading with the men all that day but made no headway, but our advice had the desired effect.

Crevasses on the left bank of the Atchafalaya May 24th, at McRae, created alarm among relief agencies, and they acted again without authority. They advised residents of towns along the right bank of the Mississippi River and Bayou LaFourche to prepare to move to refugee camps. The population of these towns aggregated some 100,000 persons. Long distance telephone calls came to me from every one of these towns, telling me that they had received instructions to get ready to be moved to the refugee camp. I told them that the flood waters would not disturb them, and that they with their belongings could remain in their homes with safety. This was followed by a special bulletin as follows:

> May 25th:—Water from the McRea crevasse is spreading with unusual rapidity, showing an enormous outflow through the crevasse. Point Coupee Parish should be evacuated immediately, and West Baton Rouge and northern portion of Iberville Parish within the next three to five days, while roads are passable and the Texas and Pacific railroad is still in operation. Suggestions have

been made that the Texas and Pacific Railroad embankment be opened to help relieve flood conditions from the McRea crevasse. Such action will not lower the flood crest either above or below that railroad, and nothing is to be gained by such action. Operation of the railroad should be continued for present evacuation and for early assistance in rehabilitation of that region after the flood. Plaquemine and Donaldsonville, and points between those places having an elevation of 20 feet or more above sea level, and towns on the banks of Bayou Lafourche from Donaldsonville to Thibodaux should not be evacuated. No town below the Southern Pacific main lines should be evacuated at present.

About 2 a. m., May 26th, a long distance telephone call came to my residence from New Iberia—The speaker said, "This is the Mayor of New Iberia, and there is a wave of water now about five miles above here and coming down this way. What is it going to do?" I reminded him that we had advised them May 21st that flood waters would overflow the banks of the Teche into New Iberia. I was afterwards told that the flood expert who had told them they would not get flood waters in New Iberia was at the Country Club in New Orleans at that time. The New Iberia people got him on the telephone and requested him to come and see the flood, which he had predicted would not reach that city.

On May 30th, Mr. Watkins, Vice-President of the Southern Pacific lines came to see me. He asked me when a stage of 10 feet would be reached on the Morgan City gauge. He stated that the law required that the draw span of their bridge at Morgan City be opened for boats going into or out of the Atchafalaya Basin whenever they signalled for an opening. The draw, if opened at a stage of 10 feet, could not be closed, and the operation of trains would have to be suspended. Mr. Watkins told me he was preparing to give notice to the public that the operation of trains over the Southern Pacific Lines out of New Orleans would be suspended on the date I set for the occurrence of the stage of 10 feet at Morgan City. I recognized that the public needed this train service, the only one operating between the east and the west, south of Saint Louis. The people along the line of this railroad, who had been forced into the second stories of their homes by the rising flood waters could not get along without this train service to supply them with food and other necessities. An emergency existed. I asked Mr. Watkins what required the opening of the draw. He said there were boats in the Atchafalaya Basin which would want to get back to the Gulf of Mexico, and the draw would have to be opened, otherwise the Railroad would be liable to a penalty which it could not stand.

This emergency called for unusual and decisive action. The suspension of train service would bring great suffering and cause starvation to many people. I advised Mr. Watkins to continue the operation of his trains on schedule until he should be forced to open the draw on the signal of a boat after the water reached 10 feet at Morgan City. I told him that the emergency was such that I would issue special advices to boats to get out of the Atchafalaya Basin at once. A special bulletin was issued as follows:

> May 30th:—The Southern Pacific bridges over the Atchafalaya River and Bayou Bouef are not considered safe for operation of draw spans with flood stages 10 feet or higher on the Morgan City gauge; these bridges will be closed to river traffic at that stage, which will be reached in the next few days. Boats which desire to leave the Atchafalaya Basin within the next three weeks should pass out through these bridges immediately.

The boats went out of the basin at once, and the Southern Pacific Railroad continued to operate its trains throughout the period of the flood.

Flood control now in effect will relieve the Lower Mississippi from such disastrous floods in the future. The general public should now know that the U. S. Engineers have devised a system of flood control which will in the future prevent the occurrence of such calamitous floods as have occurred at intervals during the last two hundred years. When we read of the disastrous floods of years gone by, and realize the great accomplishments of the U. S. Engineers in their mastery of flood control we marvel at their achievements.

(See Complimentaries in Appendix B)

CHAPTER XL.

APPRECIATED COMMENTS BY THE NEW ORLEANS ITEM-TRIBUNE

The New Orleans Item-Tribune in its issue of July 3, 1927, contained an article by Hudson Grunewald under the caption, "Doctor Cline, Unsung Hero of Big Flood". The article is reprinted here in full.

* * * *

Thousands owe their lives to him. In the eyes of Louisiana and all America his work during the recent disastrous floods in the Mississippi Valley was a work of heroism.

Heroism, the true story of which is for the first time revealed through an interview and through the study of records which disclosed certain facts in a case modestly kept in the background.

History with its Pheidippides, its Engineer Park and its Paul Revere contains no more striking example of a man through whose timely warnings thousands of lives were saved than that of Dr. I. M. Cline, chief of the weather Bureau at New Orleans, who, in forecasting the movement and intensity of the unprecedented overland flood which spread southward over the valleys of Louisiana and in warning the people in the danger zones to flee has achieved one of the greatest feats in flood forecasting on record.

The feat of Paul Revere is known to every schoolboy, Pheidippides has become legend, Engineer Park who sped down the Conemaugh valley on horseback warning the people of Johnstown to flee for their lives has gone down into history, but while no such spectacular incidents accompanied the work of Dr. Cline who silently and tirelessly toiled at maps and figures to save lives and property the name of this humble scientist of New Orleans is also destined to be prominently written in the annals of worthy deeds.

APPRECIATED COMMENT BY N. O. ITEM-TRIBUNE

During the disastrous flood epoch Dr. Cline put aside all thought but that of the vital work before him. Harnessing himself to the task he forgot all hours, all time, but worked like a trojan to study the vast and unprecedented problem so that he might issue bulletins and make forecasts and warnings so necessary for saving lives and property within the areas of threatened destruction.

Yet he refuses to talk about himself and about the valuable work that he organized and planned, giving all credit in his self effacing manner to the workers in his office. He states that his entire force worked with zeal and energy in handling the enormous amount of detail which devolved upon his office during the great disaster. He says they recognized neither Sundays nor week days, night nor day, their first thought being to serve the people who were in danger. A service which though he will not admit it, would have been impossible without his leadership and guidance and the inspiration of his own unending toil.

"Too much credit cannot be given Dr. Cline for the work he has done," recently commented Secretary Herbert Hoover in charge of flood relief and rehabilitation work in the Mississippi Valley. "His flood forecasts have been absolutely uncanny in their accuracy. He has without doubt saved the lives of thousands of people with these bulletins.

"The engineers have been aided. His reports have been so accurate that they could watch from day to day the increasing danger at points along the river. The relief workers have been aided immeasurably. They have been told exactly where to concentrate boats and trucks, what areas were most necessary to evacuate and, almost just how much water was going to rush through each crevasse as it opened."

As early as March 23 Dr. Cline began flood forecasts and warnings for the lower river below Vicksburg. These warnings were kept before the residents of the valley continually, being changed and raised as conditions in the upper valley warranted.

On Sunday, May 1, 1927, the levees along the Mississippi river in northern Louisiana had given way in several places and this presaged a great and dangerous overland flood in the lower valleys of Louisiana. Dr. Cline visualized a flood of vast proportions and on this date issued warnings for flood levels in the next two or three weeks two to four feet higher than any floods previously recorded in the history of these lower stretches of the valley in Louisiana and the people in the valleys below were warned to take every precaution.

A HIGHLY PRIZED HONOR

"THE TULANE UNIVERSITY OF LOUISIANA. Be it known—That in recognition of his exalted character, his eminent attainments in the science of meteorology, his constant devotion to truth and the welfare of society, the ADMINISTRATORS OF THE TULANE UNIVERSITY OF LOUISIANA have this day conferred upon

Isaac Monroe Cline

the degree of DOCTOR OF SCIENCE with all the rights, honors and privileges appertaining thereto. New Orleans, Louisiana, June sixth, nineteen hundred and thirty-four.

Esmond Phelps, A. B. Dinwiddie,
President of the Board. President of the University."

Tulane University had conferred this degree only three times during the 100 years since it was founded in 1834.

CHAPTER XLI.

AN INCIDENT OF THE GREAT FLOOD.

The following story is reprinted from the Shreveport "Baptist Message" of the issue of June 23, 1927. Aside from its interest, this is a story of a Mississippi River flood written in the dialect of people who live in certain sections of Southern Louisiana. It is good psychological study of these decendants of the French Acadians.

The residents of the little village had been warned, repeatedly, of their danger. A crevasse was iminent, and they were directly in its path. But we find it very easy to believe something that we wish to believe. The historic crevasse of '82 had not touched them. So all that day business went forward as usual. Talk on the streets naturally resolved about the impending crevasse. But always there was the same conclusion: There never had been, and never would be, such another crevasse as the one in '82. And the water then had only reached the outskirts of the town where one might still see, here and there, large iron spikes driven into giant cypress trees, marking the height of the water.

"Dit donc! Me, Hi tink dat faller, Doctor Cline, he's crazy! Was he here in '82? Mais, non! Me, Hi was right here. Leetle faller, den, but Hi 'ave not forgotten. My ole fadder (God bless him, he's dead now) he live right on dees house. An' water com' on de front step, but no more! All you fallers ron' way, eff you 'fraid. Me, Hi stay here, you bat!"

Two days later this same doughty doubter, who lived just on the edge of the village, was being helped down off his roof into one of the relief boats from the Red Cross camp. Perched beside him on the roof was a bedraggled Plymouth Rock hen, which clucked querulously. And inside his shirt were a dozen baby chicks.

"Mais, no, nobody here but me. Yes-ter-day Hi 'ave send my woman an' de five keed by de boat. Me, Hi stay. Already was water two feet deep in my 'ouse, but Hi tink she will not com' more 'igher. Gar! She still risin'."

"But didn't you all get Dr. Cline's warning?" asked his rescurers.

"Sure! But Hi say, 'ow he know 'ow 'igh dis water com'? Was he here in '82? Me, Hi was right here. Was leetle faller den but Hi 'ave not forgotten, My ole fadder (God bless him, he's dead now) he lives right on dees 'ouse. An' de water com' on de front step, but no more . . ." and so forth.

"Hi was down town wen de water com'. You know ware is dat cafe, by one beeg live oak tree? Always is som' peop' onder dat tree, set 'roun' an' smoke, an' drink coffee. Jus' set 'round', do nossing, all day, jus' smoke, an' drink coffee. So Hi was set onder dat tree, dat night, doing nossing, all-so. Jus' smoke an' drink coffee. Lots of keed on de banquette, my leetle Marie, too, com' by me, all dose keed walk up an' down de banquette, dragging dose 'lectric car . . ."

"'Ow is dat? You do not know wat is 'lectric car? Enfin, you teck beeg empty carton—hat-box, som'ting lak da, an' cut som 'ole' on all side, all kind pretty shape, lak hearts, stars, lak dat, you onderstan', an' you cover dose 'ole wit tissue pape', red, green, blue—all pretty color. Den you put candle inside, an' drag him on de banquette, wit a string. So. All de keed meck does 'lectric car, dis tam o' year, drag him on de banquette, c'est beau! You never see som-ting lak dat? Ware you live? Oho! Hi see!"

The "electric cars" referred to, it might be well to explain, have a very definite place in the spring schedule of games in South Louisiana, just as marbles, kites and tops have, the country over. But just as the children of this, and many other South Louisiana towns have their own system of "hop-scotch," and their own quaint vocabulary and rules of the marble ring, just so the Jack o' Lantern, sacred to Halloween, here is converted into the " 'lectric car," fashioned from a strawboard carton and colored tissue paper, making its appearance in astonishing numbers, along toward the last of May, or early in

June. It is no unusual sight to see an impromptu street parade of two or three score of these " 'lectric cars" dragged along the streets by the tiny little Creoles, some of them barely able to toddle. For perhaps a week they hold full sway. Bedtime hour, for once, is hooted at, and everybody sits up until ten o'clock, to watch the 'lectric cars. Then, all at once, just as marbles, stilts and tops have a way of doing, they all disappear overnight and are seen no more until the following summer.

"Enfin, eversing quiet, nossing doing. Till som'body say, "Hi tink Hi 'ear som' rain!" Mais, dere is no cloud! It is only de win'. Den de church bell ring, an' all de whistle begin to blow. An' som'body ron down de strit, an' 'oller, 'Crevasse! Crevasse! De water is com'! Galop vite!'

"Dit, donc! Maybe you tink was some ex-cite-ment, hah? An' jus' den, all de light go out! Pouf! Lak dat! Everware was womans cryin', keed cryin', all de peop' in de dark! Pretty bad! Was no light, only from dose 'lectric car, jus' candle, on-der-stan'? So Hi teck my leetle Marie in my arms, wit dat 'lectric car, an' Hi start for my 'ome. Marie—ma femme, on-der-stan', he was at 'ome, with four younge 'keed. An' de water was com' from dat way. Me, Hi start to cuss, but tink mebbe Hi better pray. Soon was water by my knee, den by de waist, an' all tam get deeper, an' swif'! You bat! Jus' wen Hi com' by de church, I meet my wife, calling me, com' queek! She have com' to meet us, on-der-stan', with all four keed in his arm An' den de win' blow out dat candle! Pretty bad! Beeg trobble, you bat!

"An' den—de lights com' on! Da's pretty good, hah! So. We go on de church, ware was plenty peop'—seex hondred, mabbe t'ousan', womans, keed, men, planty peop' in de church. It is high off de ground', on-der-stan', an' lots peop' go to church, dat night, never been tree-four year, mabbe. Da's funny, hah? You bat!

"So. We stay on de church, all night, all nex' day, till boats com' from the Croix Rouge, on-der-stan', teck all de peop' away. An' all dat night dose light burn. Da's pretty good, hah? Me, Hi don lak de dark. All dose womans, dose keed, afraid of de dark. An' ever tam Hi look out on dat water, an' see dose strit light still shining. Hi say, Pretty bad! But eef day was no lights, it was moch worse. You bat!

"So. De boat com' teck away all de peop'. But me, I sweem back out by my 'ouse, to see eef water is by dat front step—Nom du chein! It is by de roof, encore! An' up in de hayloft, Hi fin' dis ole hen. He 'ave hatch all dese leetle chick. My 'orse is drown', all my peeg drown, my crop—all gone! Nossing is lef', only dis ole hen, an' dese leetle chick. Pretty bad, hah? You bat!

"Mais, about dose light, Hi do not on-der-stan'. Hi tink Marie— ma femme—she on-der-stan'. She say it is one miracle. Out dere in de darkness, with dose four keed in his arm, she pray for dose light to com' back, an' dey com'. She say her prayer is answered, an' Hi tink she's right!

"Sandwiches? You bat! An' you 'ave 'ot coffee in dat bottle? Pense donc! Merci, m'seur, 'ti 'peu, 'vous plait. 'La!

"So. It was only dat the fuse blows out? An' so soon as dat is feex, we get our light from some other town? Enfin, Hi know nossing about dat—trans-mi-cion lines an' plants that in-ter-con-nect. Me, Hi know dat de light goes, an' in a few minutes we have light again. An' you say dose light come so queek from a town was more as feefty miles away? Hah! Da's fonny."

CHAPTER XLII.
COMMENTS ON TROPICAL CYCLONE OF SEPTEMBER, 1938

Newspapers friendly disposed towards the Weather Bureau consider it their duty to point out where warnings might have been issued which would have aided the public in protecting life and property. Mr. George N. Coad, one of the editors of the New Orleans Item and a close student of meteorology, has emphasized in the following article some points to which forecasters, charged with the issuance of warnings for tropical cyclones, should give special attention. Mr. Coad cites three tropical cyclones where warnings, which likewise might have advised people to move out of the danger zone in advance of the storm tide, and saved many lives. I did not assist Mr. Coad in the preparation of this article and I knew nothing of it until the article appeared in print. Mr. Coad in The New Orleans Item, in its issue of September 27, 1938, says:

Water, not wind, is the hurricane's arm of death. There are exceptions, such as the disaster of Santo Domingo, but they are few. The thing that happened last week on the southern New England coast is the rule.

A wind of 75 miles an hour is bound to do a good deal of damage, but by far the greater number of well built houses, even if greatly exposed, will stand up under even stronger winds. The splintering is done by the storm tide.

The wind itself will topple a few houses, kill some scores in their wreckage, kill a few others with flying debris or live wires, but that wall of water, taking hideous, sudden form in the night-black rain, coming on with the speed of an express train, engulfing, lifting, twisting, sweeping away houses, boats, bridges, docks, everything except the greatest breakwaters, that it is which pounds out life in a black, screaming surge.

When Isle Dernier was blasted away in 1858, when Galveston was smashed in 1900, there was excuse for meteorologists' failure to give warning. But there has been no excuse for these 20 years. There was no excuse in 1925 when Lake Okechobee swept out of its bounds and drowned thousands, no excuse in 1935 when the Atlantic sent a tremendous wave over the WPA camps on the Florida Keys and left five or six hundred men dead; no excuse last week when the Atlantic threw itself upon the New England coast.

There was no excuse for failure of the Weather bureau to warn the hundreds of persons on exposed and low places of the great probability of danger on the afternoon of September 20; no excuse to warn them on the morning of the 21st that the probability was virtually a certainty.

But no warning of an on-rushing wall of water was issued.

The Weather bureau did not consider the storm to be a hurricane on September 20, but it did so consider it on the 21st at 9:15 a. m., apparently changed its mind at 11:23 and estimated the wind as a whole gale; at 1:21 it still held to the whole gale estimate but warned that the center would cross the Long Island and Connecticut coasts in the next few hours.

The bureau cannot be blamed for misestimating the intensity of the wind. The storm had been moving along the coast for two days and ships had gotten out of its way.

There were, therefore, insufficient reports to give an estimate of the winds at the center a high degree of accuracy. The center had apparently increased its rate of movement which is generally a sign of diminishing force of the winds moving into it. The estimate that it was a whole gale, may therefore, go unchallenged.

But there still is no reason why warnings of dangerously high tides on that coast should not have been issued.

All the information needed to make such a forecast had been available to the bureau for years. In a series of publications beginning 20 years ago, Dr. Isaac Monroe Cline, formerly principal meteorologist at New Orleans, set down the relation between the movement of the center of a tropical storm and the tides upon the neighboring coasts. In his book, "**Tropical Cyclones,**" Dr. Cline consolidated all the data which he had collected as a major part of his life's work. The Weather bureau would not sponsor his book because his theory of the movement of winds in a hurricane differed from the official theory. His theory explained how storm tides are created, how they move, where they will strike. All the tide and wind records collected over a period of 30 years substantiate it.

The storm tide on the Florida Keys occurred where the Cline theory would have put it; the storm tide on the New England coast occurred where the Cline theory would have put it. Had Dr. Cline been forecasting the movement of that hurricane, some hundreds of persons, smashed and dead today, would have been alive.

The conventional theory of a hurricane is that the wind spirals inward on all sides. That theory does not explain how the tremendous storm tides, ranging as high as 15-feet, originate. If the hurricane is small in diameter, as most of them are, especially the most severe ones, then the wind is blowing parallel to the course of the storm and in the direction the storm is heading, for only a few miles. Such a short sweep of wind, even of intense force, is not enough to drive 15-foot walls of water upon exposed shores; not enough to raise waves of 50 feet in the open sea. Even if that short sweep of parallel wind were enough to raise such a wave, the theory puts even fiercer cross winds against it in the advancing right hand quadrant of the storm. If the theory were correct, certainly that water would be beaten back and slowed up.

The records gathered by Dr. Isaac M. Cline and made the basis of his theory, indicate that the wind does not spiral inward evenly around the center of the hurricane. Instead there is an immense sweep of wind in the right hand rear quadrant running almost parallel to the path of the center, a sweep that may be from 200 to 600 or 1000 miles. That is the force which raises the storm tide and drives it upon the coast a little to the right of the center of the line of movement.

The longer the sweep, the greater the storm tide for a given wind.

Now, the storm which hit New England, whether hurricane or whole gale (the bureau estimated it at hurricane strength until the day before it hit) had moved on an almost straight line for 600 miles. That immense sweep of south wind had been pushing the water up ahead of it, slowly raising the tide on the New England coast, slowly because there was to the north of the center a strong off-shore wind holding that up-heaped water back. But that mound of water was there driven by millions of horsepower of energy. The moment the center got close to shore and the off-shore wind swung around to the south, it roared upon the land like a flood from a broken dam.

That is the way hurricanes kill. That was the lesson of the Florida Keys, that was the lesson of Galveston, that was the lesson of a score of hurricanes in the past 20 years. There were all the thoroughly documented lessons of the major disasters of a quarter century on the Gulf and south Atlantic coasts to warn the forecasters that this storm, hurricane or whole gale, would, beyond any doubt whatsoever, batter the low and exposed New England coast with an immense wall of water with waves from 30 to 50 feet running over it.

But the Weather bureau at Washington did not issue any warnings of this on-rushing wall of water. The Weather bureau at New York, specifically charged with making tide forecasts, did not issue any warnings. The people along that shore, ignorant of tropical storms, read nothing alarming in their morning newspapers; nothing even in their early afternoon newspapers.

They went about their business that day until mid-afternoon when the ee-ee-ee of the wind made them realize this was no ordinary noreaster. Many still were in their houses and clubs on the beach.

Suddenly there came terrific hammer-gusts from the east, then from the southeast, trees were bending now, roofs flying.

One incident reported in the New York Times must have been duplicated scores of times.

A woman and two sons left a fashionable club at Bailey's beach. They got upon the bridge leading to the mainland. An immense wall of water rolled over them. Their automobile vanished. The debris of the club washed over them, surged back and forth upon tremendous waves.

CHAPTER XLIII.

TROPICAL CYCLONE, NORTH ATLANTIC, 1944.

ANOTHER tropical cyclone, that of September 8 to 16, 1944, traveled northward in the Atlantic Ocean, curving towards the northeast, with its center a short distance off the coast line. Dr. Charles F. Brooks, Director of the Blue Hill Meteorological Observatory, Milton, Massachusetts, in writing me on September 14, 1944, said: "Weather Bureau has broadcast a forecast of unusually high tides for New England Coast and Long Island as major hurricane heads northward. Here is your tidal discovery at work."

The United States Weather Bureau rendered an efficient service both as to storm tides and wind velocity, which saved many lives and property of great value. However, the damage resulting from the winds and storm tide was estimated at over $100,000,000. Reports in the newspapers stated that five United States vessels were lost in the storm: the Destroyer Warrington, two small Coast Guard vessels, a Mine Sweeper, and the Lightship Vineyard Sound. The loss of these vessels is surprising when the amount of information available concerning the destructive power of these storms is considered. If the cyclone had moved in across the coast as that of 1938 did the storm tide would have been much higher than that recorded.

Commanding Officers of some of the ships on duty as observers in the Gulf of Mexico during the season of tropical cyclones have conferred with me in their studies of these storms. I have advised them to keep well clear of the right hand segment of the cyclone, and to make use of the winds of least velocity found in the left hand rear quadrant where they can go up against opposing winds and approach the area of lowest barometer, so as to be able to render valuable reports without too much risk.

Time, September 25, 1944, under the heading "Science" commenting on this tropical storm contains statements which do not appear to conform to recorded facts. In commenting on the feat of two aviators who flew over the cyclone while off Chesapeake Bay and got back safely this magazine said:

> Their chief scientific observation besides its horizontal circular motion, a hurricane has strong upward air currents at its vortex and down currents at its perimeter. The plane was sucked up so steeply at the vortex that it was just like going up in an elevator.

The vortex of a cyclone is assumed to be over the area of lowest barometer at the earth's surface. Ascending air in the cyclone results from the physical and mechanical forces operating within the cyclone, and its most rapid ascent takes place over the area where precipitation is heaviest. Automatic records made during the passage of traveling cyclones, which moved in on the Gulf and South Atlantic Coasts from 1900 to 1924, show in all these cyclones that there was little or no precipitation in the rear of a line dividing the cyclone into front and rear halves. Charts made from these records show that there is a convergence of the winds of the right hand rear quadrant of the cyclone with the winds of the right hand front quadrant in the vicinity of the line dividing the cyclone into front and rear halves. This convergence causes a rapid ascent of the air in the right hand front quadrant which carries over into the front of the cyclone, but does not extend back over the vortex, the lowest barometer at the earth's surface.

The heaviest precipitation occurs about 80 miles in front of the lowest barometer and to the right of the line dividing the cyclone into right hand and left hand halves. The ascending air mass from which precipitation is falling has a higher temperature than the surrounding air at corresponding elevations, because the latent heat released as the result of condensation retards the rate of cooling of the ascending saturated air mass. Because of this the ascent and rarefaction of the air mass becomes more rapid and further intensifies the condensation of moisture and thus augments the intensity of the precipitation directly under the region where the air mass is ascending most rapidly. The aviators apparently encountered

Bronze Tablet Presented to Dr. Cline by the Southern Pacific Lines.

the rapidly ascending air mass some 80 miles in front of the lowest barometer at the earth's surface. The plane would of course be carried up rapidly in passing through this part of the cyclone. The lifting force of the air in this part of the cyclone has been observed at the earth's surface. When the plane passed outside that part of the cyclone where there were ascending currents, the reaction to the changed conditions might have given the aviators the impression that there were down currents.

Low pressure at the center of the cyclone moves towards the area of heaviest precipitation, and that leads to the conclusion that the core of the cyclone is inclined from the low barometer at the earth's surface forward into the cloud region over the area of heaviest precipitation.

Observed facts show that wind directions in the cyclone do not have "horizontal circular motion", neither do they show "strong upward currents at its vortex and down currents at its perimeter". They show that wind directions in traveling cyclones do not have a spiral inward movement around the center, but that there is systematic difference in the inclination of the winds towards the center in the different quadrants of the cyclone. Wind directions in the cyclone represent, not simply that of rotation, but the combination of translation with rotation. Observed cloud directions in the cyclone indicate that the air mass from which precipitation falls is carried forward by the general winds of the region which carries the cyclone forward and goes on in advance of the cyclone until it loses its identity.*

---

* See TROPICAL CYCLONES, Cline, The Macmillan Company, New York, 1926. Also QUARTERLY JOURNAL OF THE ROYAL METEOROLOGICAL SOCIETY, London, England, April, 1930, pages 203-205.

CHAPTER XLIV.

THE GARDEN OF EDEN AND THE ARK IN AMERICA

In this chapter the Bible story is followed as far as it goes. The purpose is to show that the legend relative to the location of the Garden of Eden is not correct.

WHILE on an inspection trip at Apalachicola, Florida, in May, 1930, I visited the shop of a cabinet maker who was reproducing antique furniture from woods which are found in the Apalachicola swamps. A block of wood attracted my attention. It struck me as being quite unusual. The cabinet maker told me that it was a piece of wood from an exceptionally rare tree. The block of wood was about two feet in diameter and of about the same length. The cabinet maker suggested that I lift the block. I thought he was joking, but I took hold of it and to my surprise it was as light as cork. The wood showed very little grain and was almost as compact as glass. He then told me that it was gopher wood, the kind of wood which the Bible tells us was used in building the Ark, and that it is not found anywhere else except in the Apalachicola swamps and in Armenia. At that time I did not give the subject further thought.

Soon after I returned to my station at New Orleans I received a mallet made from gopher wood. The mallet was carried with me and displayed when giving talks at luncheons, because the rarity of the wood and its associations added interest to the talks. I took the mallet with me to a meeting where there were about 100 men who are supposed to have some knowledge of the Bible. The mallet was held up and I said: "Gentlemen here is a piece of the wood of which Noah's Ark was built". Several smiles of credulous character were noticed which implied that they thought I had been buncoed into believing that I had a piece of the Ark. I added—"Beg your pardon, this is a piece of the kind of wood used in building the Ark. If you know the name of this wood hold up your hand." One man held up his hand. After the meeting was over I inquired of the man how he happened to remember the name of the wood. He said: "When I was a boy my father owned a farm adjoining one that was operated by a preacher who worked the farm on week days and preached on Sundays. He had a son about sixteen years old who was always up to mischief. The preacher had selected as a text for a sermon on one of the Patriarchs which started off 'And he decided to take unto himself a wife'. Saturday night after the preacher had gone to bed the boy got the Bible and pasted the leaves together so that the text read from the pulpit Sunday morning was—'And he decided to take unto himself a wife and she was built of gopher wood, 300 cubits long, 50 cubits wide, 30 cubits high, and she was pitched within and pitched without'."

A few months after receiving the mallet while on my way to the Roosevelt Hotel to give a luncheon talk it came to me in a flash that the Ark could not have been built in Armenia and then landed on Mount Ararat.

Mount Ararat in Armenia is where the Bible tells us that the Ark landed and that is about all that we are told concerning the location of the Garden of Eden and the building of the Ark. Legend through centuries has placed the Garden of Eden in Armenia. The Bible does not record any facts which would help us to locate the continent on which the Garden of Eden was situated nor the place where the Ark was built. In the first six chapters of Genesis we are told of the Creation, First Sabbath, Fall of Man, birth of Cain and Abel, wherein is mentioned the Land of Nod on the east of Eden. The first and only city mentioned in the Bible prior to the flood is Enoch, named in honor of the first son of Cain. Then follows the Genealogy of the Patriarchs, the Depravity of Man, and then the Deluge. The Garden of Eden, the Land of Nod, and the City of Enoch are the only places mentioned in the Bible prior to the time of the flood. Based on no more definite information than this the legend that the Garden of Eden was in Armenia has been accepted generally since history began.

When we relate the physical forces of nature to the Biblical record of events we conclude that the Garden of Eden could not have been located in Asia or the Ark built on that continent.

Statements in the sixth Chapter of Genesis enlighten us concerning the flood and the Ark. God instructed Noah to build an Ark of gopher wood. The Ark was to be 300 cubits long, 50 cubits wide, 30 cubits in height, and it was to be pitched without and pitched within. Writers on Biblical subjects have called attention to something which is interesting: that the proportions of the Ark are about the same as those we find in the ocean going freight steamers of today. Gopher wood is the only clue mentioned in the Bible which might identify the continent on which the Garden of Eden was located and where the Ark was built. Stories similar to the Biblical record of the flood are found either in legend or the history of nearly all lands, but such legends do not enlighten us concerning the location of the Garden of Eden or the land where Noah built the Ark, while his neighbors derided him for his foresight.

The Ark could only have been built where gopher wood grows and pitch could be obtained. The old Hebrew word, which has been translated into Gopher wood, appears but once in the Bible, and refers to a timber of a peculiarly disagreeable odor which sets it apart from other varieties of the pine family, all of which have a pleasant odor. Gopher wood of prehistoric growth, which is the same as that mentioned in the old Hebrew of the Bible, is found in northwest Florida in the Apalachicola swamps. The area in which it grows is about fifty miles square and is located on the east side of the Apalachicola River, extending from Bristol Blountstown Ferry to near the Georgia State line. Dr. Alvin W. Chapman, author of "Flora of Southern United States" (New York, 1860) discovered this tree in Florida, and classed it as the same gopher wood which is found in Armenia. Dr. Chapman named it **torreya** because of its resemblance to the pine family. When fresh cut it has a disagreeable odor, and has been given the local name of "stinkin cedar". Also because of its durability it has been called

THE GARDEN OF EDEN AND THE ARK IN AMERICA 229

"savin" wood. When the wood is dry it gives off a dust while being worked that stings like fire. The leaves have little sharp hooks on their tips that scratch badly. The yew tree and other prehistoric trees mentioned in the Bible are found growing in the Apalachicola swamps. The Florida Legislature passed a law in 1933 which had for its object the protection and preservation of the Biblical trees growing in those swamps. The Florida Forest Service has made a survey of the region in which there are trees of prehistoric growth, and has recommended that a State Park be created to protect and preserve the gopher wood, the yew tree and other Biblical trees.

Gopher wood of the kind used in the building of the Ark is found growing in only two places in the world; in Armenia and in the Apalachicola swamps in Florida. To satisfy the Biblical story the Ark had to be built where this wood is found—in one of the two places. The Bible tells us that Noah carried out the instructions which God gave him. Then it rained 40 days and 40 nights, and the flood lasted 150 days before the Ark landed. No propelling power is mentioned in connection with the Ark, hence its direction and rate of travel was determined by the elements. Between 30 degrees and 40 degrees North Latitude there is an eastward drift of the atmosphere averaging about 500 miles a day, varying with the seasons. It is this drift of the atmosphere which carries our weather changes from the west towards the east. The drift of the atmosphere would have carried the Ark towards the east at a speed which can only be speculative. However, if the Garden of Eden had been in Armenia and the Ark had been built there the drift of the atmosphere in 150 days would have carried the Ark far to the east of Mount Ararat. It appears that the Ark could have been built only with the gopher wood of Apalachicola, Florida, and pitched within and without with tar from the pine forests in the Carolinas and Georgia. Here we would have not only the eastward drift of the atmosphere, but the Gulf Stream which would aid in carrying the Ark towards Mount Ararat. The distance from South Carolina to Mount Ararat is about 8,000 miles and the rate of travel to cover this distance in 150 days would be about 55 miles a day.

When the Ark landed Noah and his family did not know that they had been transferred to another continent. They thought that they had been drifting around over the mountains of the Carolinas and Tennessee and had landed on a mountain which they had known on this continent as Mount Ararat. Prior to the flood the Mississippi River was named the Euphrates and the Ohio was named the Gihon. When the Ark landed on a mountain in Armenia they called it Mount Ararat. There was a river to the west which they thought was the Euphrates and they called it by that name, and on the north was another river which they thought was the Gihon and they called it by that name. We find in the southeastern portion of the United States today much of the same vegetable growth that is mentioned in the Bible as indigenous to the Garden of Eden. We have the serpent, the apple, the fig leaf, and other flora and fauna mentioned.

During the first three or four years after I had learned that the gopher wood is indigenous to the Apalachicola Swamps in Florida, I was under the impression that no one previously had advanced the opinion that the Garden of Eden was located in the United States and that the Ark was built on this continent. After delivering a luncheon talk one day on "Big Things in Weather" I closed with a reference to the flood and the Garden of Eden. A Methodist preacher, a reporter for one of the papers, asked me if I had seen a book Entitled, "THE GARDEN OF EDEN AND THE FLOOD". It was written by Bishop John C. Keener of the Methodist Church and published in 1901. I had neither seen nor heard of the book. While searching for Bishop Keener's book some one sent me a clipping from an old issue of the Richmond Christian Advocate giving extracts from an address by Bishop Keener on "Higher Criticism", in which the Bishop said:

> In twelve hours of this place (Charleston, South Carolina) are beds, fossil beds which contain the bones of every animal that I have ever heard of, every animal whether mentioned in geologies or natural histories, and not a few of them; for they comprise 65% of that vast deposit of phosphate of lime in the Ashley beds . . . in

THE GARDEN OF EDEN AND THE ARK IN AMERICA 231

these beds are found the bones of the megatherium, the teeth of the beaver, the horse, the Virginia deer, the gigantic shark with teeth six and one half inches long indicating the length of a body 120 feet . . . the muskrat bones and those of the o'possum, the corprolite, the ichthyosaurus, the teeth of the gigantic saurians, mastodons, the tiger, the elephant, and all those animals that live in the neighborhood of man.

I finally obtained a copy of Bishop Keener's book, and read it with great interest. He bases his conclusions on certain geological formations, such as the Ashley beds of phosphate near Charleston, South Carolina. These beds of phosphate of lime are rich in both land and marine fossils of every conceivable kind and are evenly distributed over the surface of an area more than 50 miles in diameter. The beds average six feet in depth. At the Cotton States Centennial in New Orleans in 1884, fossils of parts of 130 different kinds of animals from the Ashley Beds were on exhibition. Now, how does it happen that the bones of all these animals along with the bones of human beings are found grouped in one locality?

Bishop Keener, it appears, did not know of the growth of prehistoric trees found in the Apalachicola swamps. He assumed that the Ark was constructed of pine from the Carolinas. The Gulf Stream was the only force which he visualized as carrying the Ark to Mount Ararat. He did not consider the eastward drift of the atmosphere in these latitudes even as a contributing factor. The Gulf Stream would do its part in carrying the Ark northward to the latitude of Mount Ararat, but the eastward drift of the atmosphere would have to be considered the force that pushed the Ark across thousands of miles of water to land on another continent.

The Garden of Eden implies a place of delight, and Noah and his family must have been delighted when the Ark touched ground on Mount Ararat, for they thought they were again in Paradise. They did not know that they had been transferred to another continent. The original Garden of Eden was rediscovered by Christopher Columbus in 1492, 5,692 years after the flood had removed the original inhabitants to a different part of the world. The Garden of Eden was not only rediscovered, but it has since been redeveloped, and people today find more delight in living on this continent than in any other part of the world.

Some recent investigators have placed the date of the flood at 4,200 years before Christ. When astrology was popular, astrologists attributed the flood to a conjunction of the planets with the "watery" sign of pisces. Their calculations led them to predict another flood to take place in 1542; this prediction caused great consternation and thousands of people fled to the hill tops. President Aurial of Toulouse, France, built an ark for a flood that never came.

One of the greatest floods in the history of this country was that of the Mississippi River in 1927 when steam barges served as modern arks in transporting people to places of safety. There were floods in the Mississippi River 130 to 140 years ago which judging from legend were somewhat greater than that of 1927.

The most sudden flood in recent years occurred in August, 1929. A lofty Himalayan Glazier, which had dammed a stream for centuries, gave way and caused a flood in the valley of the Shyok, causing a rise in the water of 93 feet in 24 hours.

CHAPTER XLV.

A VOUDOU QUEEN.

VOUDOU is the name of a heathen God represented by a God-serpent through which the wishes and blessings of Voudou are communicated through a King and Queen to the members of the cult. The worship of Voudou had its origin among the Africans of the Arada Nation. Through the God-serpent Voudou is an all powerful and Supernatural Being, recognized by the cult as the creator from which all things have their origin. The supernatural powers of Voudou are exerted through a snake which is kept in a box in front of the King and Queen.

The Voudou cult was introduced into the West Indies by slaves brought from Africa. During the revolution in San Domingo in the early part of the nineteenth century, whites and Negroes escaped and sought asylum in Louisiana. Among the Negroes from San Domingo who transplanted the cult to New Orleans, a native African of the Congo Tribe known as "Dr. John" was the King and Marie Laveaux the Queen of the Voudous. The King and Queen ministered to their subjects through inspiration from the God-serpent and their tenure of office was for life. They exacted unbound confidence from their followers who were sworn to secrecy. Violation of their oath was an offense to the God-serpent, and death was the penalty. The superstition of the Negroes caused them to submit blindly, and to comply with the rules of the cult without question. The annual meetings of the Voudou cult were held in some secluded spot along Bayou Saint John during Saint John's Eve. Here each member was called before the God-serpent and required to renew the oath of allegiance to Voudou through the King and Queen.

"Dr. John", the King of the Voudous, is said to have been a fluent talker and always dressed in style. He had an office on Bayou Road near Esplanade Avenue. He posed as an herb doctor, and his office was full of bottles filled with strange concoctions, which he claimed would cure the various ills to which human beings are subject. He was adept at slight of hand, and claimed to be an astrologer and mind reader. His place was frequented not only by Negroes but by whites as well, and it is stated that white women thickly veiled were frequently seen riding up in their carriages to visit the Negro oracle. He exercised a bad influence on the superstitious of both races and his followers were numerous. Dr. John had charms and amulets which he claimed would bring good luck, and he sold them at fabulous prices. He accumulated wealth through his nefarious practices. He died at an advanced age in the late 1860's. A person called "Dr. Alexander" succeeded him as King of the Voudous.

Marie Laveaux was Queen of the Voudous during the 1830's and 1840's. The New Orleans Directory published in 1840 gives the location of her residence, 179 Saint Ann Street, which is between Burgundy and Rampart streets. She is said to have been an intelligent woman of fine physique and with a dominant bearing. As a professional hair dresser she gained entrance into the homes of the best families in New Orleans. She carried messages between clandestine lovers and arranged rendezvous for them. She was cunning, and to hoodwink the general public she invited prominent people as guests at dances in honor of Voudou. However, these pepole knew nothing of the secret conclaves held with "Dr. John" in some secluded spot known only to members of the cult. Her house was filled with charms and concoctions, which she claimed would make wishes come true whether for good or evil. Some of the drugs she sold would keep evil spirits away, and bring happiness to the possessors. Other concoctions enabled their possessors to inflict injury on those whom they disliked. There were articles which would bring sorrow and trouble to persons on whose premises they were deposited.

Marie Laveaux was consulted by politicians and office seekers who thought they were getting aid from Voudou by employing the charms she dispensed. Politicians would purchase "mascots", and sporting men would wear charms carved from the bones of human beings dug up in grave yards. Money poured into her coffers to such an extent that she was able to keep a summer home on the lake. The name Marie Laveaux does not appear in the New Orleans Directory for 1852, and it is inferred that her reign ended some time in the late 1840's.

Some time in the Sixties or Seventies the police found the secret meeting place of the Voudous. A surprise raid enabled them to capture all the members of the cult and seize their paraphernalia, including the God-serpent representing Voudou. This paralyzed the Voudou organization to such an extent that their annual secret meetings were done away with and the meetings made open to the public. However the charlatans continued their nefarious practices in an underground way.

Malvina Latour is mentioned as the queen of the Voudous in the early Eighties. No record is found of other Queens of the Voudous between the time of Marie Laveaux and Malvina Latour, but there was probably one other Queen who on account of the vigilance of the Police kept her identity hidden from the public.

The Police have been so vigilant that the worship of Voudou is not carried on openly. However the Voudou cult is probably still in existence. Some colored and some white people fear the invectives of Voudou. The Voudou high priests were unscrupulous racketeers, and they ran one of the greatest rackets of that time.

The superstitious believed that a clay figure pierced by an arrow and placed at the door of a person by an enemy or a jealous suitor would cause the violent death of that person, and the enemy would never be apprehended. In recent years I have heard intelligent persons say that "gris-gris" had been placed on their door steps as a warning that they would suffer injury to be brought about in some way by the Voudou.

Persons who visit the tomb of Malvina Latour are said to make prayers over her grave to the Voudou, seeking her intercession with the Voudou God. When their wishes are unrealized they punish the dead Voudou queen by scratching her tombstone with some metal instrumnet. That Malvina is not very successful in assisting her devout followers is evident from the manner in which her tombstone is defaced.

In the portrait of Marie Laveaux reproduced here holes in the face and body may be seen. These were probably made by supplicants who wanted to punish her for failure to get Voudou's help when requested. The dealer who bought this portrait several years ago told me that he had obtained it from some colored people. Probably because of education they had lost their faith in Voudouism, hence their reason for disposing of the picture. The portrait, however, had been pierced through until it appeared beyond restoration.

CHAPTER XLVI.

OBSERVATION ON WORLD WAR II.

PACIFISTS are indirectly responsible for World War II. Through their influence at the close of World War I, the United States Army was disbanded and the greater part of our Navy dismantled. They did not keep themselves sufficiently informed on world affairs to learn what the Prussian war-lords were doing, and opposed preparations for the defense of the United States. The pacifists apparently could not bring themselves to believe that Hitler was creating the most brutal and merciless army in the world, and that the leaders of that army were planning to conquer the world. Plans for this conquest were outlined in writings attributed to Corporal Hitler. These writings made it clear that the Prussian war-lords, imbued with the spirit of supermen with a mission to rule all Mankind, were preparing for world conquest. The United States was singled out particularly as one of the countries to be taken over,

Germany has always been dominated by the Prussian Nobility who controlled the army and through it the State. History shows that the Prussians were always a cruel and ruthless people. The kernel of the present Germany was the Mark of Brandenberg, and around this kernel present day Germany was built up by bloody, costly, and exhaustive wars in which the Prussians came out victorious, if not always in arms, then at the peace table. Prussians have been infamous for the brutality displayed in dealing with those under them and with those who do not agree with them. Their conquests through centuries have drenched Europe with rivers of blood.

Germany until World War II had not been invaded for 130 years. The war-lords always sued for peace before victorious enemies entered their country. Until recently they could tell their people that no foreign army had marched on German soil for nearly a century and a half. The Prussian war-lords outgeneraled the Allies in November, 1918. They quit before it was too late. "He who fights and runs away, lives to fight another day". They shunted the blame for World War I onto the Kaiser, and retained their own powerful organization. They were allowed to keep the nucleus of an army, and from that day forward they planned a world conquest which culminated in World War II. From the day the Treaty of Versailles was signed the Prussian Noblemen who had waged the war, whether in uniform or civilian clothes, became members of a potential army that would some day renew its effort to conquer the world. The German people failed to make reparation payments, although American and English private investors loaned them billions of dollars. They wiped out their internal debts by inflating their currency. In the meantime, the junkers and military caste shrewdly and secretly prepared for another war. They retained enough privileges to enable them to carry out their bold desgins.

The history of the Prussians should have put the Allies on their guard. The Prussians who control Germany have war and conquest instilled in their blood. The Versailles Treaty did not kill the snake, but only scotched it. Two United States statesmen, former President Taft and Senator Lodge, urged the retention of a considerable military force to prevent future wars, but could not prevail upon the Allies to include something of that sort in the treaty.

Corporal Hitler, the paper hanger, was the man the Prussian war-lords needed to aid them in carrying out their diabolical plans. Frequenting beer houses and cafes where the hoi polloi gathered to eat and drink, Hitler came in touch with the masses suffering from the effects of a severe economic depression. Hitler espoused their cause. An orator, he soon attracted attention. He made the Jew a scape goat and put on him the blame for all their sufferings and troubles. This demagogue soon gained a hold on the German populace. Like a drowning man ready to grasp at a straw, they grabbed at his specious arguments and grandiose plans for a better Germany. He soon had them believing that National Socialism was the panecea for all their ills, and that the extermination of the Jews was necessary. The demagogue became a demi-god.

The Prussian war-lords saw in the popular Hitler a dupe, whom they could use to sway the middle and lower classes. They silently and shrewdly helped to make Hitler the leader and ruler of Germany. Under his fanatic leadership they could bring about a war of world conquest. If by any chance the war went against Germany, the fault could be charged against him and the German people instead of the Prussian war-lords.

Hitler, as is the case with most neurotics, was cunning. When on trial for conspiring against the German Government he pointed his finger of accusation at the judges who were trying him as the real conspirators. Imprisonment did not weaken his belief that he was to become the leader of the German people to conquer the world. Although sentenced to a long prison term he was released in less than one year. It is evident that the powerful influence of the junkers was behind him.

After Hitler's release from prison money flowed into the Nationalistic Social Party treasury. (Something similar has been going on in the United States. Money is now flowing into the treasury of the Congress for Industrial Organization.) Who planned and carried through the drive, which in July, 1932, enabled the Nazis to poll 13,700,000 votes and get control of Germany? What were the Prussian Noblemen doing behind the scenes? Who aided Hitler in writing the books and dissertations which bear his name? If Germany had won World War II the facts might have come out and we might have known the answers, but perhaps now we shall never know.

When, if ever, the questions are answered, it will be found that the Prussian military caste were the "ward leaders" who brought about the large Nazi vote, that they furnished the money, and put Hitler in power, and that they even collaborated in writing the books that are said to have been written by the Fuehrer. Through Hitler the German people were trained in the brutality of the Prussians, instructed in cruel methods of warfare and lead to commit crimes against civilized peoples such as had never been known in modern times.

Notwithstanding the preparations for war which were being carried forward so extensively in Germany, the Japanese "incidental" war in Asia, and the avowed intentions of the Prussian war-lords as expressed through their mouthpiece, Corporal Hitler, the United States Government made no counter military preparations. Powerful armies were being organized, which had for their avowed object the conquest of the United States along with the balance of the world. Our statesmen failed to sense the danger.

Claims have been made that Congress was asked to appropriate money to fortify some of the Pacific Islands. Congressional records do not show that such recommendations were made. A bill was introduced to provide for the improvement of a harbor on the Island of Guam, but no appropriation was made. Congress acted wisely in refusing to make appropriations for this harbor, because it would have helped the Japanese and not the United States. Guam was taken by the Japanese on December 11, 1941. We can be thankful that there were not more American boys there to be taken to Japan as prisoners of war than the Japanese found. Our small army in the Philippines was left to meet its fate without any provision for its assistance. The several thousand men on that faraway post were expendable.

On July 10, 1944, J. Carlton Ward, Jr., President of the Fairchild Engine and Aircraft Corporation, who was head of a commission in France to advise on aircraft production, in a testimony before a Congressional committee, made some startling revelations. He testified that "diplomatic sources" had permitted him to see while in Paris before France fell, a detailed plan for the invasion of the United States prepared by Hitler in 1940 to be put into effect after England had been conquered. The plan provided for a feint through Newfoundland, while the real attack on the United States would be made by way of Mexico with tanks and other armored equipment. Mr. Ward said Germany not only had plans for the military invasion but for the economic domination of the World. Our Government did not seem to realize that the situation was serious. Mr. Ward brought out the fact that in 1939, Gen. H.

H. Arnold asked Congress for 5,500 planes, and that money was appropriated for only 375 military planes.

After World War II was started, the Axis powers carried it forward with a swiftness unparalleled in history. The United States was soon forced into the war, although it had only a small army and a moderate-sized navy. Fortunately there were able and progressive officers in the United States Army and Navy who had kept abreast of military affairs. Our general staffs prepared to organize and lead armies and navies to victory. They moved in swift and efficient action. Patriotic citizens of the United States stood ready to fight in defense of the principles for which the founders of this Republic had fought and died. Free enterprise responded promptly, and manufacturing plants of every kind were converted to the manufacture of the implements of war with surprising rapidity.

Labor as a whole rendered good service. However, there have been strikes in the ranks of organized labor every year since the United States entered the war, which interfered with and delayed the manufacture of war materials. These labor strikes one year numbered more than 5,000. It is true that the number of men on strike usually was not large, but such strikes interfered with the work of thousands who were not on strike. This made the cumulative loss of production as the result of strikes enormous. When laborers in time of war stop work for personal gain and because of trivial grievances they should be considered as disloyal and as guilty of treason as the soldier who throws down his arms and refuses to fight on the field of battle. Many of those who went out on strike were not strikers at heart, but feared to attempt to continue working because of the dreaded and brutal picket line. Laborers have the same right to organize for their betterment as business men, but a man should not be required to join a labor union in order to obtain employment in any industry where needed. Such a requirement is undemocratic and is contrary to the principles of the Constitution of the United States.

Congress should pass just and fair laws for the protection of both labor and business in their rights under the Constitution. Unless such laws are passed this country will drift into

Fascism and this nation will meet the fate that overtook Italy and Germany. Initiative in the individual should be encouraged, and there should be an incentive for the ambitious. A man should be compensated on the basis of the character of work he performs, and not on the basis of the time utilized in doing a piece of work.

Nearly six months before the Pearl Harbor disaster, on December 7, 1941, I received a post card from Honorable Burton K. Wheeler, United States Senator, in which he requested me to write the President of the United States, asking that he keep the United States out of war. I knew enough about the intentions of the Prussian war-lords to realize that war was coming, regardless of any action the United States Government might take in an effort to avoid the catastrophe. The following letters explain themselves and show my stand on the matter:

New Orleans, La., July 25, 1941.

Hon. Burton K. Wheeler,
U. S. Senator,
Senate Office Building,
Washington, D. C.

Sir:

This acknowledges receipt of your card asking me to write President Roosevelt requesting him to keep us out of the war.

In this connection I will say that I am 100% with President Roosevelt in his all-out aid to Great Britain and Russia in this extreme emergency. They are fighting our battle for the preservation of freedom and we must help them win, and if necessary to that end I favor our active entry into the war.

Adolph Hitler said to Hermann Rauschning, who quotes him in his book "Voice of Destruction", on the Nazi world revolution doctrine pertaining to America as follows: "America is permanently on the brink of revolution. It will be a simple matter for me to produce unrest and revolution in the United States, so that these gentry will have their hands full with their own efforts. I shall long have had relations with the men who will form a new government—a government to suit me. We shall find such men .... in every country. We shall not need to bribe them. They will come of their own accord. Ambition and delusions, party squabbles and self seeking arrogance will drive them ....."

Is it possible that the citizens of the United States will allow this to happen as it happened in France? Will they allow the freedom which our forefathers bequeathed us to be taken from us as outlined by Hitler?

Are there in the United States the counterparts of Petain and Laval who will aid in turning the Government over to the Nazis?

Respectfully,

Isaac Monroe Cline.

P. S. I have served my country in dangerous posts and although 80 years old am ready for service in the defense of our freedom.

I. M. C.

* * *

New Orleans, La., July 25, 1941.

The Honorable Franklin D. Roosevelt,
President of the United States of America,
White House,
Washington, D. C.

Honorable Mr. President:

Referring to the inclosed card from Senator Burton K. Wheeler, I want to tell you that I am 100% with you in your all-out aid to Great Britain and Russia in this extreme emergency. They are fighting our battle for the preservation of freedom and we must help them win, and if necessary to that end I favor our active entry into the war.

I spent 53½ years in the United States Weather Service. During the Spanish-American War, 1898-99, I spent much of the time in Yucatan establishing, in cooperation with the Mexican Meteorological Service, meteorological stations for use in giving hurricane warnings to the United States fleet in their operations in the Caribbean Sea. There was a severe yellow fever epidemic in Yucatan while I was there but it did not deter me for I realized that if death overtook me I would be happy dying in the cause of freedom.

I volunteered for service as meteorologist in the Fire and Flame Division, United States Army, in World War I. The Chief of the Weather Bureau told me that I was needed in the position I occupied and that a man who could not fill my position here could serve as meteorologist in the Fire and Flame Division in France.

I have given a special contribution to my country in a book TROPICAL CYCLONES, published by the Macmillan Company, New York, which has material not only an aid to areas subjected to hurricanes but that can be used and made of great value to the Army and Navy.

I shall lend what aid I can in the purchase of Defense Bonds.

Very respectfully,

Isaac Monroe Cline.

Two well groomed men, aged between 25 and 30 years, came into my store in the summer of 1940 and commenced talking to me in the German language. I let them talk for a few minutes and then told them that I did not have any knowledge of the German language. I informed them that my ancestors on both sides came to this country more than 200 years ago, and that I am thoroughly American. They departed without further ceremony. These men were evidently Fifth Columnists or propagandists. Pacifism in this country was the result of German Propaganda, carefully planted and concealed in order to make it appear that the origin was with the citizens of the United States. The Prussian war-lords thought this would prevent the United States from building up a strong army and navy and thus our nation could be easily conquered.

Display of military power will always be respected by beligerent nations. If the United States had built up a large army and navy instead of wasting money on such activities as the W.P.A., which instead of strengthening, weakened the morale of the people, World War II would not have occurred. The United States with a large army and navy could have made a gesture to the Prussian war-lords that would have stopped them. Brutal persons are cowardly when confronted with a power as great or greater than their own.

Hitler and the German people stand charged along with the Prussian war-lords with many murders and brutal crimes committed. A striking fact is that the word **Prussian** was seldom mentioned in the newspapers and rarely heard over the radio in connection with their brutalities. How have the Prussians, the leaders in cruelty and crime, succeeded in covering up their identity as organizers and leaders of this brutal war? Punishment to suit their crimes will be difficult to mete out. The Prussians, as leaders of the army, can not be punished too severely. The instigators and perpetrators of crime, whether Prussians or Germans, should be executed and their families scattered to the four winds of the earth. The entailed estates of the Prussian Noblemen should be given to the serfs who have been their slaves for centuries. Such action will break up the junker aristocracy, and discourage the formation of brutal armies in the future. The severest punishment possible should be meted out to the Japanese.

Great Britain, by the heroic acts of her people in the early part of the war, saved civilization from being destroyed. If the German armies had over run Great Britain they would have invaded America before the shovels of the W.P.A. could have been changed into implements of war. However, the Prussian war-lords decided that they would first wipe Russia off the map in four months and then cross the Atlantic to finish the job in America. Great Britain and Russia were aided by the United States, and Russia turned the tide against the German armies. The United States was forced into the war by the treachery of the Japanese on December 7, 1941, under conditions which then existing will prove to have been of world benefit. It put the citizens of the United States on their metal in the future. And civilization though badly crippled has been saved. It is hoped that the wounds of this cruel war will be healed in time.

Selective Service has found approximately five million young men of draft age incapacitated for military service because of physical or mental defects. This is appalling. Such defects in most cases can be cured early in life. Every boy, when he reaches the age of seventeen years, should be taken into the United States Army for two years. Those with physical or mental deficiences should be placed on limited duty under the care of the excellent medical service of the United States Army. In this way the standard of manhood of the United States can be greatly improved and the country will always be ready for war. By being ready, war will then be prevented.

The atomic bomb is the most destructive weapon ever made for use in war. It was secretly developed during World War II, and its supporters were not subjected to criticism and persecution which often happens to those who are progressive thinkers. A striking instance of the consequences of such persecution is illustrated in the following: William Mitchel who, by natural ability and hard study, rose from the ranks of Private to Brigadier General in the United States Army foresaw the mastery of the airplane as an implement of war. In his books, OUR AIR SERVICE, 1921, WINGED DEFENSE, 1925, and other publications he dared to express advanced ideas which his superiors could not comprehend. Ranking officers in the army, not competent to judge, had him court-martialed and forced out of the United States Army in 1926. Twenty years later he has been vindicated. Congress in 1946 conferred upon the late "Billy" Mitchell the medal of honor posthumously and the posthumous rank of Major General in recognition of his progressive thinking in connection with the use of the airplane as a weapon of war.

Who will devise means for combating the atomic bomb? Some one will accomplish that feat.

CHAPTER XLVII.

REFLECTIONS

INDUSTRY and social contacts contribute towards keeping a person in good mental and physical condition. Idleness not only weakens the body but the mind deteriorates along with the body. The utilization of recreation time in buying and restoring works of art, in studying porcelains, glass, and allied subjects, not only refreshed me mentally, but gave me a knowledge both of what made an article a work of art and its comparative value. These studies broadened my viewpoint and enabled me to carry forward my regular duties in the weather service more efficiently than otherwise would have been the case.

After 53½ years in the weather service of the United States I was not satisfied when retired at the age of 75 years, to lead an idle, listless life—perhaps to die in a few years, as so many persons do after retirement. The art knowledge which I had acquired opened a field in which I could occupy my time in a pleasant and profitable way, and at the same time keep myself in good mental and physical condition. On December 31, 1935, I was separated from the weather service, and on January 1, 1936, "The Art House—Dr. Cline" opened for business. When I was relegated to the shelf as too old to work the knowledge I had acquired in my recreation time was used in carrying on a business. This business has been conducted without the employment of any other person. I have allotted to myself a task that does not bring fatigue, but refreshes me and keeps me in good condition. When my time is not occupied entertaining visitors, I clean and restore paintings and repair the frames of paintings.

Thousands of friends (I have been dubbed the millionaire of friends) have given me their support socially and in a business way. People from all parts of the world have visited my store. People who are interested in art objects are pleasant and agreeable conversationalists. During the nine years I have been in business I have had no worries, my physical condition has improved, and at the age of 84 years my mind is as clear as it was at the age of 50. My three daughters married good men, and an annuity provides for the necessities of life, so that I have no worries. I have four grandchildren and three great grandchildren. Two of my grandsons are in the armed services, fighting for the freedom which our forefathers won for us, viz: free enterprise, free speech, freedom to work where and when we please, without the dictation of minority organizations. Time has brought much of joy and much of sorrow, but the joy has greatly exceeded the sorrow. If I had my life to live over and could plan it, I would not change things in any way, for I am satisfied with my life as it has been. My objectives in life have been accomplished.

Reading of fiction has held very little of interest for me. My reading in the main, has been of books on subjects connected in some way with the work on which I was engaged or to my hobbies. I selected books which contained matter that would aid me in my work. Newspapers always claimed my attention, and the information which they carried kept me abreast of what was going on in the world. Franklin wisely said "Dos't thou love life, then do not squander time for that's the stuff life's made of." Time is a fortune given by the Creator to every human being, the value of which depends upon its utilization. It is free for all from the time of their advent into this world until they pass to the great and unknown beyond. Time is the foundation upon which the people, in a country of free enterprise, have the privilege of building careers for themselves. In every walk of life the degree of success achieved is determined by the utilization of time.

Time lost can never be recovered, and this should be written in flaming letters everywhere, so that all may read and profit thereby. One thing which has impressed me most in life is the waste of time. The use made of time leaves its imprint and remains as an influence through future generations. When death claims us we may think we have finished our work in this world, but we are mistaken if we think such a thing.

Our acts and deeds are imprinted on other minds, and, influencing the actions of those with whom we come in contact, these influences are passed on from generation to generation. Thus the manner in which one utilizes time leaves its lasting influence on society, though as individuals we may have been forgotten.

Young and growing children should be impressed with the fact that time is of great value and should not be wasted on trivial things. The trite saying—"As the twig's bent the tree's inclined" is as true in the formation of the mind and habits in young and growing children as in the formation of the twig which grows into the sturdy tree. The training of the child determines the character and achievements of the man or woman. The tree as it grows may be twisted sharply for brief periods by hurricane winds of the cyclones, but, unless destroyed, the tree after the passage of the winds returns to the shape in which it had grown from the bent twig. The human being may be tossed hither and thither in the maelstrom of strife, war, pestilence, storms and floods, but the teachings received in childhood continue to assert themselves through all vicissitudes of life.

Recreation is any diversion which brings rest after toil. Since time is the stuff life is made of, recreation time should be utilized to better fit the individual for his vocation and at the same time to benefit mankind. Physical development is essential to mental development, and sports which develop the body and lengthen life should be encouraged. However, the present day tendency is to waste time in sports which neither contribute to the development of the body nor the mind, but on the contrary often result in permanent injury and impairs the efficiency of the individual through life. Such sports should be discouraged.

Children should early in life be taught to think of some vocation and to prepare themselves especially for that line of work. The child will grow in the direction in which its mind is trained. My father told me when I was a small boy that he intended to give me an education so that I could do something different from farming. My ancestors on both sides, with the exception of a school-teacher now and then, were tillers of the soil. I look back on them with pride, because the States was built and agriculture supports it today. There is no more important and honorable calling than that of the farmer.

Services are always compensated for in proportion to their value. There is no greater reward than knowing that one's services have benefited mankind. For 53½ years the public always had first call on my time. Generally speaking, my day has been divided into three parts, eight hours for work, eight hours for recreation, and eight hours for sleep. However, when situations presented themselves where the public interest could be served, time was no consideration. Sixteen hours out of the twenty-four were often devoted to serving the public, and when the danger was grave I have worked twenty to thirty-six continuous hours. Overtime pay was neither received nor expected. The benefits which the public received were sufficient compensation. "Topics and Personnel" of November 1942 (United States Weather Bureau), in emphasizing the importance of confidence and promptness in meteorological service cited a feature of my work for the information of all Weather Bureau personnel. (See Appendix C.) Some of the results of my recreation hobbies have been not only of temporary but of permanent value to the public as shown in preceding chapters.

Advancing years call for certain precautions. Worries should be avoided. Bridges should not be crossed in the imagination for many of the imaginary troubles we contemplate will never occur.

The fable of the tortoise and the hare carries an important lesson that very few persons heed. It teaches that steady methodical work accomplishes the object in view. The winning of the race over a stretch of time is not to the swiftest but to the steady systematic worker who plans his campaign so that he can accomplish what lies ahead of him without fatigue. The systematic worker performs as efficiently at the end as when he started, whereas the one who hurries does less efficient work towards the end and often fails in accomplishment. Haste often delays and causes loss of time. It is not making good use of time but burns up time which would yield good results when used in steady methodical ways. When a person hurries he burns up energy more rapidly than the mechanism of the body can supply new energy. When a person hurries fatigue comes quickly. Fatigue not only weakens the body but impairs the efficiency of the mind. Much more is accomplished by slow steady application than by hurrying. In the great crises in which I have always been successful, I never allowed myself to hurry, and, when I was through, the work was as efficiently performed at the end as that at the beginning. I always safeguarded myself against being overcome by fatigue.

As persons grow older the energy expended in performing work is not replenished with the same rapidity as was the case in earlier years. Proper attention to diet, habits of living, and the manner in which people perform their duties will enable them in their 70's and 80's to exhibit the mental and physical fitness of persons of 45 to 60 and still retain the alertness, grace, and charm of middle age.

STORMS, FLOODS AND SUNSHINE
PART II.
CHARACTERISTICS OF TROPICAL CYCLONES
(Hurricanes, Typhoons and Baguios)

Summarized from TROPICAL CYCLONES by Isaac Monroe Cline, 1926.

Important material which, heretofore, the reader and student had to get from the tables and illustrations, in my book TROPICAL CYCLONES, has been brought into the text in this publication.

FOREWORD

THE JOURNAL OF THE ROYAL METEOROLOGICAL SOCIETY (Great Britain) April, 1930, gives TROPICAL CYCLONES a three page review from which the following extracts are taken. "The book constitutes a notable advance in the collection and representation with precise data with regard to the tropical cyclones * * *. A new method has been devised for charting meteorological data during the passage of a cyclone * * * . Studies of cyclones have previously been based mainly upon synoptic charts which do not enable wind directions, and velocities, the distribution of precipitation and cloud movements to be shown in relation to the cyclonic center. The method devised by Dr. Cline remedies such deficiencies. * * *. No such complete data as to the conditions obtaining in tropical cyclones have ever before been assembled, and Dr. Cline is to be congratulated upon the persistence and enthusiasm with which he has carried out his self appointed task. The book will prove of the greatst value to meteorologists in their search for the ultimate causes of cyclones, both of the tropical and temperate varieties. It should also be of great value as an aid in the forecasting the destructiveness of a cyclone in particular localities."

Isaac Monroe Cline
October 13, 1950

STORMS, FLOODS AND SUNSHINE

PART II.

CHARACTERISTICS OF TROPICAL CYCLONES

INTRODUCTION

WHERE TROPICAL CYCLONES OCCUR

Tropical cyclonic storms occur in six widely separated parts of the world, four north of the equator and two south of the equator. Provincial names are given them in some places.

NORTH OF THE EQUATOR

1. The West Indies, the Caribbean Sea, the Gulf of Mexico, and the southern portions of the North Atlantic Ocean. In these regions tropical cyclones are usually referred to as "hurricanes". The name "hurricane" comes from a Carib word which signifies high winds, but has no connection with the circulation or movements of the winds in the cyclonic area. Cyclones are most frequent in these regions in July, August, and September, but have occurred in all months from June to November inclusive. Tropical cyclones, after they leave the tropics, have been known to travel, and retain their identity, over long distances through the temperate regions. Some times hurricane winds and storm tides caused by these cyclones bring destruction to the New England coast.

2. On the western coast of Mexico and thence westwards towards the Hawaiian Islands in the North Pacific Ocean. Here they are called "tropical cyclones". They occur occasionally from June to October inclusive.

3. Over the China Sea, the Philippine Islands, and Japan, in the North Pacific Ocean. In the China Sea and Japan they are called "typhoons", while in the Philippines they are sometimes called "typhoons" but more often "Baguios". The word "typhoon" is derived from a Chinese word meaning a strong veering wind and has about the same meaning as the word "hurricane". "Baguio" is said to be of Malay origin and is synonymous with "typhoon". In these regions cyclones occur from May to December, and are most frequent in July, August, and September.

4. On each side of India in the Bay of Bengal and in the Arabian Sea and in the regions of the North Indian Ocean. In these regions they are called "tropical cyclones". In the Bay of Bengal they have occurred in all months of the year except February, and there are two periods of high frequency, one in May and the other in October and November. In the Arabian Sea and the North Indian Ocean they have occurred in all months of the year except February and August. There are two periods of high frequency, one April, May, and June, and the other October and November.

SOUTH OF THE EQUATOR

5. To the east of Madagascar in the vicinity of the Islands of Mauritius and Reunion, and thence over Keeling Island to the northwest coast of Australia, in the South Indian Ocean. Here they are called "cyclones". They have occurred in these regions in all months except July and August, and are most frequent from December to March inclusive, with the greatest monthly number in January.

6. From Australia to Paumato Islands in the South Pacific Ocean; here they were formerly called "typhoons" or "willie-willies", but now are generally referred to as "cyclones". They have occurred in this region in all months of the year with the greatest frequency from January to March inclusive.

There are no records of a tropical cyclone having been encountered in the South Atlantic Ocean.

* * * *

All tropical cyclones belong to the same family. These storms have different names in some regions of their occurrence, but they are all of the same formation and have the same general characteristics. The proper designation, regardless of where they occur, is "tropical cyclone". In the nothern hemisphere tropical cyclones develop between 10 degrees and 20 degrees north latitude. Cyclones can not develop in the immediate vicinity of the equator because there is so little deflection of the air in that region, from the earth's rotation, that air flowing into any depression would remain radial, and the depression would soon fill up and die out without cyclonic devel-

CHARACTERISTICS OF TROPICAL CYCLONES

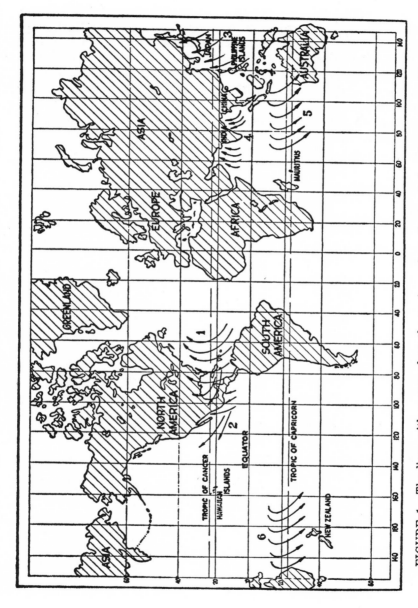

FIGURE 1.—The lines with arrows show the regions where the tropical cyclones occur. The curvature of the lines with arrow show the directions which the cyclones follow. The arrow points in the direction of travel.

opment. North of the equator these cyclones first have a westerly direction with a speed of 8 to 12 miles per hour, the velocity of the trade winds or the winds of the region in which they develop. The greater number of these cyclones curve to the northward in latitudes 25 degrees to 30 degrees and after recurving slowly move off with the winds of the region towards the northeast. South of the equator they curve towards the southeast in the vicinity of the Tropic of Capricorn with the velocity of travel the same as that of the winds of the regions through which they pass.

CHARACTERISTICS OF TROPICAL CYCLONES
(Hurricanes, Typhoons, and Baguios)

Summarized from TROPICAL CYCLONES, by Isaac Monroe Cline, 1926

(Figures in parenthesis refer to Bibliography on page 337)

The Galveston, Texas, cyclone (hurricane) of September 8, 1900, when 6,000 persons were drowned, not killed by the winds, convinced me that the accepted theories giving the directions of the winds in the different parts of the cyclones as they move around the center are not correct and are misleading. The illustrations in text books and other publications show the wind directions moving spirally inward around the cyclonic center with considerable uniformity in all quadrants of the cyclone.

A striking example of such an illustration is found on page 207, Figure 100, AN ILLUSTRATED OUTLINE OF WEATHER SCIENCE, 1945, Pittman Publishing Company, New York and Chicago. The isobars are correct, but the wind directions are widely different from those which actually occur in the different parts of the cyclone, and are misleading to students and navigators of the sea and air. William Ferrel 75 years ago showed by mathematical deductions that the winds in a tropical cyclone are different from those shown in illustrations referred to. Wind directions moving spirally inward around the cyclonic center could not have developed the storm tide which wrought such destruction and wiped out so many lives in a few hours. I have spent much time during the last

50 years studying the contributions of others to our knowledge of the characteristics of tropical cyclones. A brief summary of what others had contributed to this subject is given in my address as President of the American Meteorological Society, in 1934, (Appendix A, page 318 of this book.)

William Ferrel in 1871 showed that neither the radial theory of Espy nor the gyratory theory of Redfield as modified by others could be correct (38). In 1885 he published a contribution (41) on this subject which should have convinced all meteorologists, who had the inclination to go seriously into the study of cyclones, that the accepted theory of the nearly uniform spiral inward movement of the winds around the cyclonic center is not correct. The reason Ferrel's great contribution to the study of cyclones was not accepted can only be attributed to the petty jealousy of fellow scientists and investigators in the study of meteorology.

NEED FOR UNIFORM NOMENCLATURE

In studying the writings of other investigators, in different parts of the world where tropical cyclones occur, I have found it difficult to compare descriptions and the conclusions arrived at by various writers, because reference was generally made to the points of the compass in describing the characteristics of the weather phenomena in the different quadrants of the cyclone. They did not seem to realize that cyclones wherever they occur are all of the same family. In a cyclone traveling in a westerly direction, the conditions on the north side of the cyclone would be entirely different from those on the north side of a cyclone traveling in an easterly direction. This rule applies in all parts of the world.

Since cyclones travel towards different points of the compass, I have adopted a nomenclature that will aid in the study of tropical cyclones, and make descriptions of cyclones in any part of the world comparable with those anywhere else. The designations "Front", "Rear", "Right" and "Left" in referring to the halves of a cyclone, convey a clear meaning as to the half of the cyclone referred to in discussing its characteristics. The quadrants are designated as "Right front qua-

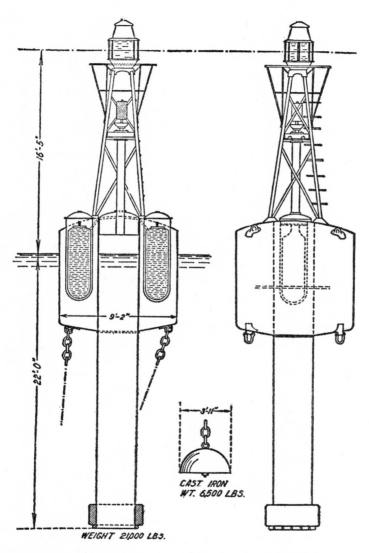

FIGURE 2—Design of Gas and Whistling Buoys anchored at Trinity Shoals, Louisiana, Galveston (Texas) Bar, off Jetties and at Heald Bank, Texas.

drant." "Right rear quadrant," "Left front quadrant," and "Left rear quadrant." These terms convey a more intelligent meaning as regards a particular part of the cyclone than when referred to by the points of the compass.

IMPORTANCE OF TOPOGRAPHY

Topographic conditions are of the greatest importance in the study of cyclones (hurricanes, typhoons, and baguios). Meteorological stations in the coast regions of the Gulf of Mexico and the South Atlantic States offer more favorable conditions for the study of these cyclones than in any other part of the world in which they occur. Here the land is comparatively level for a good distance inland from the coast line. There are no mountains to distort the isobars and break up the cyclonic circulation, as is the case in all other regions where these destructive atmospheric disturbances occur, whether named hurricanes, typhoons, or baguios. Here the tropical cyclone, as it approaches the coast and then as it moves inland for several miles, retains its tropical characteristics. Furthermore, in these regions the storm tide builds up in the front and to the right of the line along which the center of the cyclone is advancing for some distance inland from the coast, sometimes for more than 100 miles, after crossing the coast line. Observations made at stations in this water area must closely approach conditions which characterized the cyclone while over large bodies of water. (See Figure 3.)

STORM TIDES INVITE STUDY

Forecast work for the Gulf area, commencing in 1900, led me to observe that the rise in the storm tide on the coast begins soon after the cyclone enters the Gulf of Mexico, and that the storm tide rises only in the immediate front of and in the right front quadrant of the cyclone. The storm tide covers the entire right front quadrant of the cyclone, and without exception the storm tide occurs in all cyclones in the same position on the coast as related to the center of the cyclone and the direction in which it is traveling. I recognized that this information could be used to great advantage in forecasting the extent, intensity and the locality where the cyclone would be most dangerous.

FIGURE 3—Height of storm tide above mean Gulf level caused by tropical cyclone September 29, 1915, southeast Louisiana, and on the coasts of Mississippi, Alabama and Northwest Florida. This is Figure 5, Appendix, TROPICAL CYCLONES, Cline, The Macmillan Company, 1926.

After studying this subject for several years I requested authority from the Chief of the U. S. Weather Bureau to prepare a paper on this subject for publication. He told me that the forecasters and the scientific staff in his office said that my theory could not be substantiated. Fortunately I had collected from the U. S. Engineers and the U. S. Coast and Geodetic Survey automatic tide records made during every tropical cyclone that had moved in on the Gulf and South Atlantic coasts since 1900. I showed Prof. Marvin some of my charts illustrating the action of the storm tide as the cyclone approached and moved inland on the coast. He was convinced that I had made a discovery of importance and he personally authorized me to proceed with the study. This prevented any one else from taking up that line of research. He further instructed me to show the physical forces in the cyclone which produce these destructive storm tides. This problem made it necessary for me to show pictorially that the accepted theories of the spiral inward movements of the winds around the cyclonic center, with considerable uniformity, are not correct.

METHODS HERETOFORE USED DO NOT GIVE CORRECT RESULTS

Previous studies of cyclones have been based mainly on synoptic weather charts. These do not represent the direction of the winds, the location of the precipitation and other weather phenomena in their correct relation to the center of and the direction in which the cyclone is traveling. On the weather maps the points of observation are widely separated, the observations are taken at the same time and do not give the weather elements as they change with the travel of the cyclone; therefore, we do not get a true picture from the synoptic chart of the changes taking place as the cyclone travels forward.

Observations from ships at sea have been of great value in studying the meteorology of the oceans and in making forecasts. These observations, notwithstanding their limitations, are of great value in forecasting the movement of tropical cyclones. However, it is a well known fact that ships avoid hurricanes, and consequently observations made by ships in a hurricane are few in number and widely scattered. The location of a ship's position can not be accurately determined because of the cloudiness attending the cyclone. Masters of ships who have gone through tropical cyclones have told me that when they have succeeded in correctly locating their position, after having been through a tropical cyclone, they often found that they were forced 50 or more miles from their estimated position.

Captain Lecky in WRINKLES ON PRACTICAL NAVIGATION, says:

"Owing to the fact that a ship is not a fixed observatory, and that the determination of the wind force on a steamship is far from easy, confusion can only result from efforts at a measure of the angle of indraught by means of observations taken on board ships under the influences of a hurricane when there are other matters to be safeguarded". He also says: "Hurricanes in low latitudes are of small radius and great intensity, an unknown sea surface and lack of ships position by observation are quite common on such occasions, and consequently conclusions based on such uncertain data are worse than useless—they are misleading."

I was closely associated with marine meteorological work for 47 years, and met the masters of numerous ships which followed courses that took them through the regions visited by tropical cyclones. Many ship captains have confirmed what Captain Lecky says. Therefore, it is clear that studies of tropical cyclones must be based on observations made at stations on islands and along coast lines. Stations on Islands are widely separated hence observations at coastal stations must be used in these studies.

INTEGRATION METHOD USED IN CHARTING DATA IN CYCLONES

I spent much time working out a method by which the weather elements can be shown on a chart in their direct relations to the center of a traveling cyclone; and to show what is taking place as the cyclone advances. I found that this could be accomplished by the use of the integration method. This is the equivalent of greatly multiplying the number of observations as the cyclone advances. This method enables the entry of the weather phenomena on a chart for each station in the cyclonic area as related to the center and movement of the cyclone as it travels forward (See TROPICAL CYCLONES, Cline, 1926, pages 30-34). The weather elements entered on the chart are: the wind and cloud directions at the hour, and the wind velocity and precipitation recorded during the hour ended at that time. These entries are commenced for each station as it enters the front of the cyclone, and are continued for each hour until the station passes out through the rear of the cyclone. The wind directions charted in this way show the wind paths in the cyclone which can not be shown on synoptic weather charts. Charted in this way the wind directions in the different parts of the cyclone are shown to be widely different from the spiral inward circulation shown in illustrations in text books and on many charts. We find that the wind directions when charted in their relations to the center and direction of travel of the cyclone fully account for the destructive storm tides which are recorded in all cyclones only directly in front of and mainly to the right of the projected line along which the center of the cyclone is traveling.

The integration method was first used in charting the data in two tropical cyclones which were used in my first study of the relations of storm tides to the center and movement of tropical cyclones in the Gulf of Mexico. These charts showed the winds in the right rear quadrant of the cyclone blowing continuously in the same general direction as that in which the cyclone was traveling, and this showed the physical forces in the cyclone which cause the storm tides on the coast attending tropical cyclones. (See United States Monthly Weather Review, March, 1920, 48; 127-142). The forecasters and scientific Staff of the Weather Bureau claimed that the results shown on my charts of the two cyclones were accidental and could not be taken at their face value. Their stand caused me to write my book TROPICAL CYCLONES in which all the data recorded in tropical cyclones that had moved in on these coasts, 1900 to 1924 inclusive, are charted and illustrated. This complete work was published so that any who question my findings have the data handy for use in making investigations of their own. It is now 23 years since the publication of TROPICAL CYCLONES and no one has proved my findings wrong. Please read the review of TROPICAL CYCLONES by the Journal of the Royal Meteorological Society of Great Britain, pages 184-188 this book.

In this connection Sir Napier Shaw, the distinguished British meteorologist, in a publication dated 1922, (64, 65) emphasizes what I brought out pictorially in my study and charting observations by the integration method, published in the United States Weather Review, March 1920. In referring to a discussion he had with Lord Rayleigh relative to wind directions in tropical cyclones, he says: "I recognized that in looking for pictorial evidence of the fact (which should have been evident to any one during the last 50 years) that, if the revolving fluid were carried along by a current of air, the winds would represent, not simply the rotation, but the combination of translation with rotation."

PERSONAL EXPERIENCES IN TROPICAL CYCLONES

During my residence of 60 years on the Gulf of Mexico coast twelve tropical cyclones have moved inland over the place where I resided. This has enabled me to observe personally the characteristics of these destructive atmospheric disturbances from the center outward in every direction. These personal experiences have enabled close-up observations, and I have studied them carefully. In addition to personal contacts with these cyclones I have charted the data recorded by automatic recording instruments for each observation station which was in a tropical cyclone for every tropical cyclone that has moved inland on the Mexican Gulf coast and the South Atlantic coast, as far north as Wilmington, North Carolina, for the period, 1900 to 1924 inclusive.

BAROMETRIC PRESSURE IN TROPICAL CYCLONES
(Storm Tides and Currents)

Pressure distribution in tropical cyclones is an important feature in determining the characteristics of the cyclone. The lowest barometer readings near the center of important cyclones and other phenomena attending them are given for a few cyclones. Galveston, Texas, September 8, 1900; the lowest barometer was 28.48 inches at 8:20 p. m. and this was 20 miles to the right of the line followed by the center of the cyclone. The storm tide was 15 feet at Galveston from 8:00 p. m. till 9:00 p. m. of the 8th. The storm tide in the right front quadrant of the cyclone covered the country to a considerable depth for more than 50 miles inland from the coast line. The storm tide was not noticeable 50 miles to the left of the line followed by the center of the cyclone.

TROPICAL CYCLONE AUGUST 13-17, 1915

In the cyclone which moved inland over Galveston and Houston, Texas, August 17, 1915, the lowest barometer at Galveston, Texas, 45 miles to the right of the line followed by the center of cyclone, was 28.64 inches at 2:00 a. m. of the 17th; at Houston, Texas, 60 miles inland from the coast and 15 miles to the right of the line followed by the center of the cyclone, the barometer was 28.20 inches at 5:25 a. m. of the 17th, and this indicates that there was little or no change in the intensity of the cyclone after moving inland on the Gulf Coast and traversing a distance of some 60 miles. The storm tide at Galveston was 12 feet in open water at 3:00 a. m. of the 17th and was banked up to 13 feet at La Porte at the head of Galveston Bay. The land in the right front quadrant of the cyclone was covered with storm tide for some 50 miles inland from the coast line. To the left of the point on the coast where the center of the cyclone moved inland there was no storm tide.

POWERFUL CURRENTS ON RIGHT OF PATH OF CENTER

Powerful currents were developed 20 to 100 miles to the right of the line followed by the center of the cyclone of August 13-17, 1915.

Trinity Shoals gas and whistling buoy, weighing 21,000 pounds, anchored in 42 feet of water, with 6,500 pounds sinker and 252 feet of anchor chain weighing 3,520 pounds (total 31,020 pounds) was carried between 8 and 10 miles west of its permanent location, 29 degress and 07 minutes North Latitude and 92 degrees and 15 minutes West Longitude (See Fig. 2). This buoy was anchored 100 miles to the right of the line followed by the center of the cyclone.

Galveston Bar gas and whistling buoy (Same as Fig. 2) anchored at end of Galveston Jetties in 36 feet of water was carried 4½ to 5 miles in a southwesterly direction from its proper position. This buoy was about 20 miles to the right of the line followed by the center of the cyclone.

A buoy of the same class with the same moorings (See Fig. 2) in 42 feet of water located on Heald Bank, 28 miles off the entrance to Galveston Bay, was not moved from its proper location, but its lights were extinguished. This buoy was a short distance to the left of the line followed by the center of the cyclone. This demonstrates that the storm currents did not run across the front of this cyclone. It is evident that the storm currents went on over the jetties and the land into Galveston Bay, building up a storm tide as far north as La Porte of 13 feet, and covering a large area of land in the right front quadrant of the cyclone.

This is the first instance where we have had an opportunity to get reports on the action of currents attending a tropical cyclone. The movement of the buoys mentioned above are proof that powerful currents are developed in the right half of tropical cyclones. The more powerful currents are some distance to the right of the line followed by the center of the cyclone.

TROPICAL CYCLONE, SEPTEMBER 29-30, 1915

Burrwood, Louisiana, was about 50 miles to the right of the line along which the center of the cyclone traveled. The high wind velocities recorded at Burrwood indicate that it was in the severest part of the cyclone. The lowest barometer at Burrwood was 29.00 inches at 2:00 p. m. September 29th. At New Orleans, Louisiana, the lowest barometer was 28.11 at 5:50 p. m. September 29th, about four hours after the lowest barometer at Burrwood was recorded. New Orleans was only six miles to the right of center of the calm area of the cyclone, which was recorded distinctly at Tulane University. The barometer remained stationary at Tulane University, reading 28.10 inches from 5:30 p. m. till 6:00 p. m., 30 minutes, and a dead calm prevailed during this time. The calm was about 8 miles in diameter. The storm tide at Burrwood, in open water, was 5.8 feet and at Grand Isle, 9.0 feet. The storm tide in the right front quadrant covered southeastern Louisiana with water 5 to 10 feet deep and 9 feet to 13 feet in Lake Pontchartrain. (See Figure 3). It was 9 feet to 12 feet along the Mississippi coast and was 4 feet as far to the right as Pensacola, Florida. This cyclone was over water to a great extent until its center passed north of New Orleans, and it appears to have had about the same intensity when its center passed New Orleans that it had when it moved in on the coast line. (See Figure 4). There were no light buoys or other objects to show the currents to the right of the path of the center of the cyclone but the storm tide built up shows that the currents were in operation.

TROPICAL CYCLONE, SEPTEMBER 9-15, 1919

When it comes to the study of cyclones this is one of the most important. There was nothing to indicate the character and movement of this cyclone after it moved through the Florida Straits until a short time before it moved inland below Corpus Christi, Texas, except the storm tide. Weather observation stations along the coast were too far distant from the center of the cyclone to give any idea as to the extent, severity, or direction of movement of the cyclone. The storm tides along the coast told the full story about the cyclone to one who had studied the storm tides and what they indicate.

The lowest barometer at Key West, Florida, 20 miles to the right of the line along which the cyclone was traveling, was 28.81 inches at 12 midnight of the 9th. We have no tide records in that vicinity but storm swells sent out from the right rear quadrant caused a rise in the storm tide at Burrwood, at the mouth of Mississippi River, of 0.3 of a foot at 8:00 a. m. September 11th. The storm tide continued to rise at Burrwood, until 10:00 p. m. September 13th when it was 3.6 feet. Burrwood was about 125 miles to the right of the line followed by the center of the cyclone. At New Orleans, 200 miles to the right of the line followed by the center of

CHARACTERISTICS OF TROPICAL CYCLONES 269

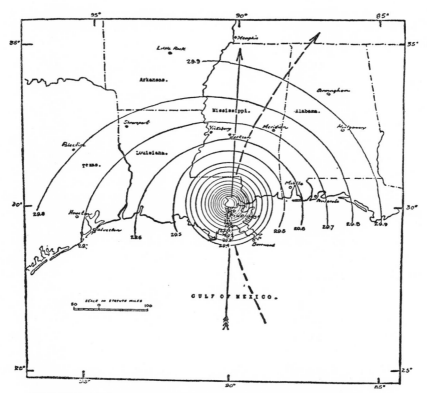

FIGURE 4—Barometric gradients in tropical cyclone, September 29, 1915, 8:00 p. m., 75th Meridian time. Isobars are drawn for each .10th of an inch of the barometer. Broken line is path followed by center of cyclone. The solid line shows the direction in which the cyclone was moving at 8:00 p. m. This cyclone after having traveled nearly 100 miles after crossing the coast had not diminished in intensity. See other illustrations of this cyclone in book TROPICAL CYCLONES, Cline, The Macmillan Company, 1926.

the cyclone the barometer was in the low 29.70s from noon of the 12th till 8:00 p. m. of the 12th; the lowest barometer was 29.71 inches at 5:00 p. m. of the 12th. The barometer 200 miles from the center did not give any information about the cyclone, but the storm tide 125 miles from the path of the center told that the cyclone was intense and moving westward in the Gulf of Mexico.

Galveston was 150 miles to the right of the line followed by the center of the cyclone. The lowest barometer at Galveston was 29.60 inches at 7:00 a. m. September 14th and the highest storm tide was 7.6 feet in open water at 3:00 a. m. of the same date. The barometer did not give us any information worth while but the storm tide indicated a powerful and large cyclone traveling westward through the Gulf of Mexico.

Corpus Christi, Texas, was 45 miles to the right of the point on the coast where the center of the cyclone moved inland. The barometer readings at Corpus Christi, the station nearest to the center of the cyclone when it moved inland, did not give any guidance until the death-dealing storm tide was overrunning the coast region and leaving destruction in its wake. The barometer at Corpus Christi at 3:00 p. m. September 13th was 29.73 inches, and at 3:00 a. m. of the 14th was 29.56 inches, a fall of .17 of an inch in 12 hours. The barometer then fell more rapidly and was 28.65 at 3:00 p. m. of the 14th when the center of the cyclone moved inland. No deductions of a definite nature concerning the intensity and direction of travel of the cyclone could be made from the barometer readings.

Storm tides along the coast, however, carried a more definite message, telling of the intensity and movement of the cyclone. The storm tide had commenced showing a rise at Aransas pass (Corpus Christi, Texas) nearly thirty-six hours before the cyclone reached that part of the Texas coast and continued rising steadily until it reached 11.5 feet at 10:00 p. m. September 14th. The storm tide banked up to 16 feet at Corpus Christi from 4:00 p. m. till 6:00 p. m. September 14th. The currents developed in the right half of the cyclone

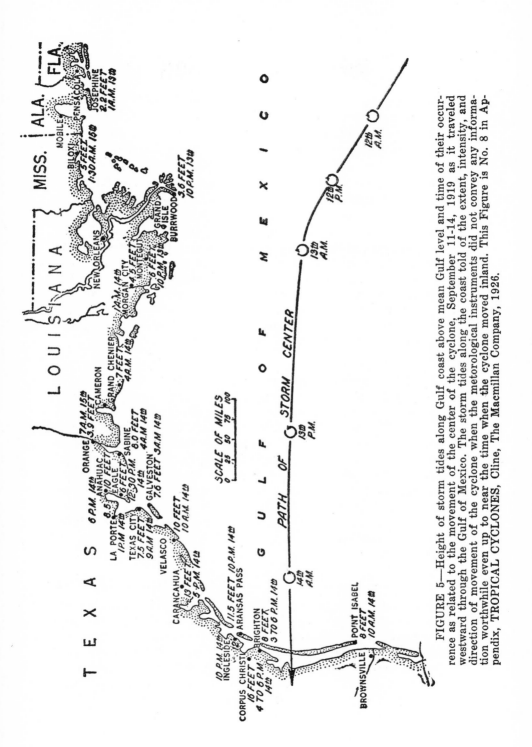

FIGURE 5—Height of storm tides along Gulf coast above mean Gulf level and time of their occurrence as related to the movement of the center of the cyclone, September 11-14, 1919 as it traveled westward through the Gulf of Mexico. The storm tides along the coast told of the extent, intensity, and direction of movement of the cyclone when the meteorological instruments did not convey any information worthwhile even up to near the time when the cyclone moved inland. This Figure is No. 8 in Appendix, TROPICAL CYCLONES, Cline, The Macmillan Company, 1926.

met some obstruction on this coast which forced the water across the front of the cyclone and gave a storm tide of 8 feet at Point Isabel, Texas, at 10:00 a. m. on the 14th some six or eight hours before the center of the cyclone moved across the coast and six hours before the highest storm tide was recorded at Corpus Christi. Point Isabel is about 80 miles to the left of the point on the coast where the center of the cyclone moved inland. This is the only instance in the 19 cyclones studied in which the storm tide showed up to the left of the point where the cyclone moved inland. Topographic conditions along that part of the Texas coast prevented the storm tide and currents from running inland and building up a storm tide on land. The currents when they encountered resistance were forced across the front of the cyclone and caused a storm tide along the coast to the left of the place where the center of the cyclone moved inland.

This cyclone furnishes a good example for use in the study of the depth of the water along the coast caused by the storm tide in a cyclone which was a considerable distance from the coast line. The storm tide along the coast from Pensacola, Florida to Point Isabel, Texas, is shown in Figure 5. The rise in the storm tide in bays and estuarys lags behind the rise on the coast. At Morgan City, Louisiana, the storm tide was 7 feet at 4:00 a. m. on the 14th, six hours after the highest had occurred on the nearby coast. At the head of Galveston Bay, La Porte, Texas, the highest tide 8.5 feet was at 1:00 p. m. on the 14th, some ten hours after the highest water in the open Gulf at Galveston, Texas, had been recorded.

POWERFUL CURRENTS IN THE RIGHT HALF OF THE CYCLONE

Powerful currents were developed by the storm swells created by the high winds of the right rear quadrant and sent forward through the right front quadrant of the cyclone. Trinity Shoals gas and whistling buoy described in Figure 2, anchored in 42 feet of water, 100 miles to the right of the line followed by the center of the cyclone, was carried 2½ miles to the westward of its permanent location.

Galveston Bar gas and whistling buoy (see Figure 2), 175 miles to the right of the line followed by the center of the cyclone was carried 1½ miles to the southwest. The currents here were apparently influenced by the Galveston Jetties and deflected somewhat to the left towards the center of the cyclone.

Aransas Pass gas and whistling buoy, anchored in 42 feet of water, Latitude 27 degrees and 50 minutes N., Longitude 92 degrees and 02 minutes W., was carried 5 miles nearly parallel to and slightly towards the coast (this shows that the currents were being influenced and forced to turn across the front of the cyclone). This buoy weighs 8,000 pounds, the anchor 5,000 pounds, and 252 feet of chain weighing 3,528 pounds, total weight, 16,528 pounds. This buoy was 75 miles to the right of the line followed by the center of the cyclone.

The destructive force of storm swells is illustrated in this cyclone. At Sabine Bank Lighthouse, 15 miles off Sabine Pass and 125 miles to the right of the line followed by the center of the cyclone, cast iron plates ⅝ of an inch thick, 27 feet above mean sea level water were bent in and crushed up by the storm swells.*

NO OTHER WRITERS SO FAR AS WE COUD FIND HAVE REPORTED SUCH CURRENTS

None of the authorities, writing about tropical cyclones in other localities where they occur, report any thing to compare with the action of currents observed in the tropical cyclones of August 15-18, 1915, and September 8-15, 1919, in the Gulf of Mexico.

At Port Elizabeth, South Africa, blocks of rubble stone weighing 100 to 150 pounds in 22 feet of water, were not disturbed by waves 15 to 29 feet high. At Columbo, stones weighing three tons, in water 10 to 14 feet deep, were disturbed by waves not exceeding 15 feet in height. At Algiers, Africa,

---

* The information relative to the movement of buoys and damage to light house was obtained from the U. S. Superintendent of Light Houses for the Gulf District.

stones 33 to 38 feet below the water were moved by waves or currents. Elliott (97) says: "It is probable that the strength of a cyclone advancing to the northward coast of the bay could be roughly ascertained by the strength of the westerly set at the head of the bay. Data are however wanting to test this."

The most destructive element in a tropical cyclone is the breaking wave. A cubic yard of water weighs about 1,500 pounds, and the waves driven by the wind, break on the coast with the water moving forward at a speed of many feet per second, and cause destruction which exceeds description.

In all of the cyclones studied, the isobars, lines of equal barometric pressure are nearly circular outward to the isobar of 29.40 inches. (See Figure 4). The filling up of the cyclone commences at the center with the isobars around the center disappearing first and the remainder of the isobars gradually disappearing as the cyclone decreases in intensity. The swells and storm tides follow the same pattern in all the cyclones studied.

GENERAL WINDS CARRY CYCLONES FORWARD

Meteorological students are generally agreed that the winds of the region carry the cyclone forward, and that the cyclone travels with a speed equal to the velocity of the prevailing winds of the region over which the cyclone is traveling.

R. P. Vinito Vines, S. J. (54): In an early contribution to the subject states that the winds in the cyclones he studied showed a great deviation towards the center; "not only at a distance from the vortex but even in its vicinity."

W. Doberck (55): THE LAW OF STORMS IN THE EASTERN SEAS, 1904; states that the prevailing winds of the region carry the cyclone forward and influence the rotary winds and states: "Less than half a mile up in the air the incurvature of the winds towards the center disappears in the average of the quadrants, but it still blows in towards the center in the rear. It is really the wind at this altitude that carries the typhoon along, for late in the autumn there are typhoons every year that move along against the northeast moonsoon, but we know it is at times very shallow and there is southeast wind above it. These typhoons disappear sometimes suddenly when the northeast monsoon increases in depth and intensity."

Rev. Jose Algue, S.J., Manila, Philippines. (53) Gives studies of detailed observations at one station (Manila) for six cyclones in his valuable contribution to the study of cyclones (typhoons) of the far East. He says: "The wind directions will be less convergent on the front side owing to the progressive movement of the cyclone, so that they will more approach the circular form than is the case in the rear. The direction of the wind on the two sides of the track undergo, owing to the direction of the track and the intensity of the progressive movement, changes which are all the greater, the greater the velocity of the cyclone".

Prof. W. J. Humphreys (73), says: "Synoptic weather charts therefore, show instantaneous wind directions, but not wind paths. This is because the storm condition itself is moving forward-moving, indeed, with a velocity nearly always comparable to and at times even faster than, that of the lower winds themselves."

Prof. Charles F. Marvin (74) Says: "It is well known that "Highs" and "Lows" travel at considerable velocities in definite directions therefore the actual direction of the winds over the ground will be those compounded from (1) the motions appropriate to the system of curved isobars and (2) the motions of the system as a whole."

Sir Napier Shaw (61, 62, 63, 64, 65). Referring to tropical cyclones, says: "They begin by traveling with the velocity of the normal current in the region of their origin, about 10 miles per hour." Shaw gives us a vast amount of valuable information on tropical cyclones which the reader will find in the publications following the above numbers.

Dr. Oliver Fassig (60): "The prevailing direction of the wind throughout the year in Porto Rico is between east and southeast".

Maxwell Hall (67) Makes a brief but important contribution to the study of tropical cyclones in the West Indies. He says: "But in most tropical hurricanes the incurvature is least in front and greatest in the rear." * * * "Over the Caribbean Sea there is a constant drift which blows over Jamaica from the east-southeast during the hurricane season of August, September and October * * * it was assumed that the constant drift flows at the rate of 8 miles per hour".

TRADE WINDS AND TEXAS MONSOON CARRY CYCLONE FORWARD

The air current in which the cyclone is carried forward in the West Indies and the United States, and which determines the direction of travel is the prevailing winds, at the earth's surface and for some distance above, in the region through which the cyclone travels. These are (a) the trade winds over the Caribbean Sea mostly from east to southeast, (b) then in the Gulf region the summer monsoonal winds, (103) a continuation of the trade winds, coming in on the Texas coast from the southeast and south and finally curving into the winds of the middle latitudes moving from the west towards the east.

The trade winds in the Caribbean Sea have average velocities over Porto Rico of 10 to 12 miles per hour (60) from east to southeast; and 8 miles per hour over Jamaica (67) from the south-southeast during the tropical cyclone season. Summer monsoonal winds come in on the Texas coast with velocities of 10 to 16 miles per hour from the southeast and south. The monsoonal winds move northward and come under the influence of the prevailing winds of the latitude and curve eastward. The depth of the prevailing air current in which the cyclone travels after crossing the coast line is shown by the depth of the monsoonal winds as far north as Groesbeck, Texas, and Broken Arrow, Oklahoma (69, 70) where they often prevail during the summer months to a height of more than two miles.

The trade winds merging with the monsoonal winds over the Gulf of Mexico, and the monsoon merging with the winds

CHARACTERISTICS OF TROPICAL CYCLONES 277

of the middle latitudes (98) form an almost continuous air current with a curve similar to a parabola, and tropical cyclones make a similar curve modified by conditions prevailing at the time of their occurrence. Occasionally abnormal air movements carry the cyclone westward.

During the months of June, July, and August about 70% of the tropical cyclones observed in the Leeward Islands and the Caribbean Sea have moved well westward in the Gulf of Mexico before recurving eastward. During September, when the prevailing winds of the Caribbean are from the east half of the time, and from the southeast half of the time, about 50% of the cyclones moved well westward into the Gulf of Mexico before recurving and the other 50% curved to the eastward without going so far to the westward. During October, when the prevailing winds over Porto Rico are from the south one-fourth of the time, only 25% of the cyclones observed in the Leeward Islands and the Caribbean Sea have moved west of Longitude 85 degrees before recurving. The prevailing winds in Jamaica from June till October do not vary greatly from those in Porto Rico. It is interesting to note how closely the changes in the paths of the cyclones follow the changes in the prevailing wind directions in the Caribbean regions as the season advances.

The direction of movement of cirrus clouds over a wide area indicate the direction of movement of the general air current at their level. In the front and the rear of the larger cyclones the cirrus clouds moved forward mainly in the same general direction as that in which the cyclones were traveling. This is shown by the charted directions of the cirrus clouds for six of the larger cyclones. (See Topical Cyclones, Cline, Figure 48). The directions of the cirrus clouds in these cyclones as recorded indicate that the general air currents in which the larger cyclones are traveling extend up to the cirrus level and that these air currents carry the cyclones forward. Cirrus clouds observed in connection with the larger cyclones indicate that the depth of the air currents which carry the larger cyclones forward is at least five miles, while the width of this air current is, as indicated by the width of the area covered by the cirrus belt, at times more than 1,000 miles.

In the smaller cyclones, those with a diameter less than 450 miles, the cirrus clouds follow their prevailing direction from the west towards the east and move on over the top of the cyclone. This indicates that the air currents in which the smaller cyclones travel are not as deep as in the larger cyclones.

The courses followed by tropical cyclones and their rate of travel are governed mainly by air currents in the cloud region, and therefore the position of high pressure areas sending out their anticyclonic circulation may materially influence the winds in neighboring cyclones and the direction in which the cyclone continues to travel. In this connection reference is made to Prof. Marvin's discussion of steady motions of winds within and between high and low pressure areas, where he shows the relations of the winds to the gradients. (74).

Sometimes high pressure areas augment the air current in which the cyclone is traveling, and on the other hand they may directly oppose and overcome this air current; in the latter case the cyclone ceases to advance, becomes stationary and fills up. The abnormal movements of upper air currents in determining the direction of the movement of cyclones is shown by the abnormal course followed by the tropical cyclone of September, 1919, (99) which has been carefully studied by Weightman.

WINDS AS RELATED TO THE CENTER OF TRAVELING CYCLONES

The wind directions at the earth's surface in traveling cyclones result from a combination (a) of the forces (modified by resistance encountered) acting on the inflow of air towards the cyclonic center, with (b) the air current which is carrying the cyclone forward. Thus we have a combination of rotation with translation. The integration method which I have used in charting the data in the cyclone as it travels forward, enables us to show the paths of observed wind directions in the different parts of the cyclonic area as related to

CHARACTERISTICS OF TROPICAL CYCLONES 279

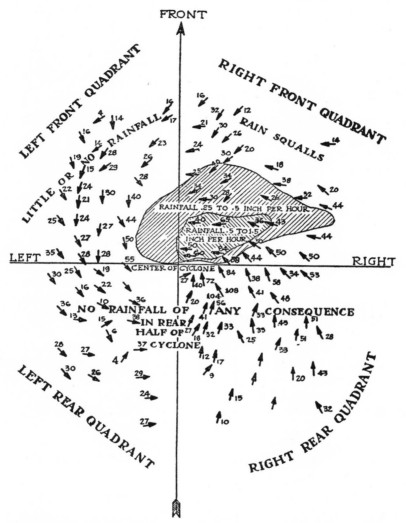

FIGURE 6—Ground plan of tropical cyclone. Made by combining the actual wind directions and velocities and precipitation in four traveling tropical cyclones having diameters or more than 450 miles, on one chart. These are the actual recorded wind directions and velocities in the different parts of the cyclone. The shaded areas are hourly amounts of rainfall from automatic records made during the passage of the cyclone.

The wind directions and velocities in this chart are from Figure 42 and the hourly rainfall is from Figure 43 TROPICAL CYCLONES, Cline, The Macmillan Company, 1926.

the traveling center of the cyclone; this had not been done previous to the publication of my book TROPICAL CYCLONES, in 1926. No one has attempted to prove that the integration method as I have used it is not scientifically sound. In my studies of tropical cyclones which moved inland on the coasts of the Gulf of Mexico and the South Atlantic States during the 25 years, 1900 to 1924 inclusive, charted by the integration method show that the directions of the winds as related to the center of the cyclone, in the different quadrants, are the same in all cyclones that continue to travel. The actual wind directions in the different quadrants as related to the cyclonic center are shown pictorially on Figure 6. The following descriptions represent the actual wind paths as they move around the center of a traveling cyclone.

In the right front quadrant of the cyclone the winds are not spirally inward towards the center of the cyclone. In addition to the forces mentioned in the preceding paragraph, the wind directions in the front of the cyclone are materially influenced by the winds from the right rear quadrant of the cyclone as will be brought out in the discussion of the winds of that quadrant. The cyclonic winds in the right front quadrant have directions almost across the projected line along which the center of the cyclone is moving. After crossing into the left front quadrant the winds are somewhat inclined towards the center of the cyclone, the winds then move around the center into the left rear quadrant where they take up a course across the line of advance of the cyclonic center and merge with the winds of the right rear quadrant, or we might say are sucked into the winds of the right rear quadrant by the high wind velocities of that area.

In the right rear quadrant, the prevailing winds that carry the cyclone forward combine with and deflect the cyclonic winds to the right, and the result is that the winds of the right rear quadrant have the same general direction as that in which the center of the cyclone is traveling; and they continue in that direction as long as the cyclone is traveling forward. The winds in this quadrant of the cyclone, blowing continuously in the same general direction as that in which

CHARACTERISTICS OF TROPICAL CYCLONES

the cyclone is traveling, form a continuous air stream at the earth's surface in the right rear quadrant of the cyclone (Fig. 7.) This air stream in some cyclones is 400 to 800 miles in length and 250 to 400 miles in width in the right rear quadrant of the cyclone. The length of the air stream depends on the size and intensity of the cyclone. These winds with their high velocities 40 to 100 miles or more per hour run into the winds of the right front quadrant which have a high velocity, and we have a convergence of these winds which spreads over the greater portion of the right front quadrant. This convergence not only causes the air to ascend but deflects the winds of the right front quadrant so that they do not curve in towards the center of the cyclone. We have observed that in the right front quadrant the winds blow in frequent, narrow, confused, and sudden powerful gusts lasting but a fraction of a second. These gusts are from varying directions, showing much turbulence. They have a much greater velocity than that shown by the greatest single mile velocity, and have ascending action as shown by their lifting power on objects in that part of the cyclone. In the larger cyclones these gusts are noticeable 100 to 150 miles, and in the smaller cyclones 50 to 100 miles, in front of the line dividing the cyclone into front and rear halves. We have here convection on a large scale in the right front quadrant of the cyclone. This convergence extends over the right front quadrant and the part it plays in the cause and distribution of precipitation in the cyclonic area will be discussed under that heading.

THE TURBINE PRINCIPLES IN CYCLONES

A study of the directions of the winds in the different parts of the tropical cyclone leads to the conclusion that the mechanical and physical forces operating in the cyclone are produced by a resultant cyclonic circulation which gives the impression of a great and powerful atmospheric turbine. In figure 7, the small arrows represent the actual paths of the winds in the cyclone, and the figures the hourly sustained velocities of the winds where the entries are made; in the right rear quadrant it will be seen that the small arrows nearly all point in the same general direction as that in which

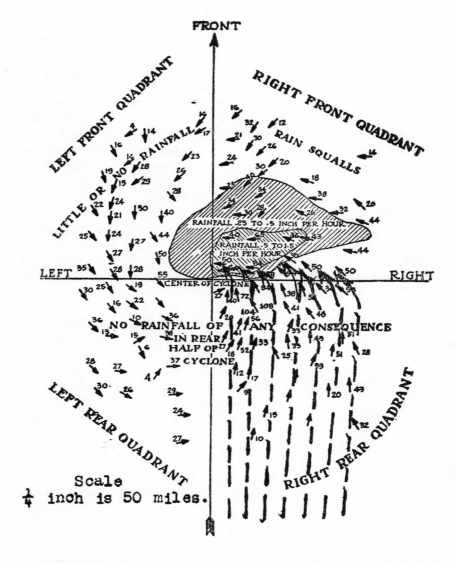

FIGURE 7—The air stream which furnishes the material that generates a horsepower shown by the energy of the cyclone so great that it is beyond the estimation of man. This air stream also sends out large swells that run forward far in advance of the cyclone, announce its approach and then cause destructive storm tides in the right hand front of the cyclone.

the cyclone is traveling. This shows the existence of a vast air stream in that part of the cyclone moving in the same general direction as that in which the cyclone is traveling, with wind velocities of 40 to 100 miles per hour. This air stream is emphasized by the larger broken arrows showing how the air stream runs forward and converges with the winds of the right front quadrant of the cyclone. This air stream furnishes the material from which the energy is produced that keeps the cyclone in operation; this energy is released directly in front of the center and over the right front quadrant of the cyclone. This will be brought out more clearly in discussing the causes and distribution of precipitation.

CLOUDS IN CYCLONES

The direction of the lower clouds show that there is no systematic incurvature of the winds in front of the cyclone at their level. The records we have the most complete available of the alto-cumulus and alto-stratus clouds show that the wind directions at their level in front of the cyclone are generally inclined slightly outward from the center of the cyclone.

The cirrus clouds recorded in the cyclones show that there is no outflow of air radiating in all directions over the top of the cyclone as would be required to support the theories to that effect; on the contrary the movement of the cirrus clouds show that the air at their level moves on over the cyclonic circulation in the large cyclones with the air current which carries the cyclone forward. Over the smaller cyclones the cirrus clouds follow their prevailing directions from the west towards the east. The cirrus are not influenced by the cyclonic circulation in the smaller cyclones.

The air carried into the cyclonic area, after precipitating its moisture and releasing its energy, is carried out from the front of the cyclone by the prevailing winds of the region in which the cyclone is traveling.

A striking feature of the larger traveling tropical cyclones is, that over the right rear quadrant, the winds at all elevations up to and including the cirrus level move forward in the same general direction as that in which the cyclone is traveling. The charts showing the direction of the winds at the earth's surface, the direction of the lower clouds, the intermediate clouds, in all cyclones, and the cirrus clouds in the larger cyclones, over the right rear quadrant, all have the same general direction as that in which the cyclone is traveling.

CYCLONES THAT STOPPED TRAVELING

There were three cyclones, in our study, which encountered opposing forces, after moving inland, that stopped their forward movement and they soon died out. The winds in these cyclones assumed more of the theoretical movement around the cyclonic center, than is found in cyclones which continued to travel, except that the winds in the right rear quadrant did not show much incurvature towards the center of the cyclone.

The precipitation occurred in all parts of these cyclones. The heaviest rains shifted position and at times fell in the rear part of the cyclone. The lowest barometer shifted positions to the locality where the heaviest rain was recorded.

WIND VELOCITIES IN RIGHT AND LEFT HALVES

The winds that carry the cyclone forward increase the velocities of the winds in the right half and reduce the velocities in the left hand half of the cyclone. The momentum of the air current in which the cyclone is traveling makes the difference in the velocities of the wind of the right and left halves of the cyclone greater than the rate of travel of the cyclone would indicate. In the larger cyclones the greatest wind velocities occur about 50 miles in the rear of the line dividing the cyclone into the front and rear halves and 40 to 50 miles to the right of the path followed by the center of the cyclones. In the smaller cyclones the highest wind velocities occupy the same relative position to the center of the cyclone. These features make the right half of the cyclone much more dangerous, both from the winds and the storm tide, than the left of the cyclone. In fact hurricane winds are seldom recorded in the left half of the cyclone. In our studies we have not noted hurricane winds to the left of the line dividing the cyclone into right and left halves.

WINDS IN REAR OF CALM CENTER IN TROPICAL CYCLONES

Some writers on tropical cyclones state that the winds rise to hurricane force after the passage of the calm center. We have found the reverse to be the case in every tropical cyclone we have charted. Wind velocities diminish rapidly at stations passing out in the rear of or near the calm center of cyclones: For example:

Galveston, Texas, September 8-9, 1900, the center of the cyclone passed 20 miles to the left of Galveston at 8:30 p. m. About three hours later the wreckage on which we had been floating settled on the ground as a result of the fall in the storm tide. The wind was nothing like as strong at the time we went aground as it was before the passage of the center of the cyclone. The anemometer blew away two hours before the passage of the cyclone center, therefore we have no record of the wind velocities. Furthermore, if hurricane winds had been blowing in the rear of the cyclone, the direction was off the Gulf of Mexico and this would have kept up a high tide; on the contrary the tide fell rapidly after the center of the cyclone passed. The tide at the time of the passage of the cyclone center at 8:30 p. m. was 15 feet, and at 11:30 p. m. was 8 feet, a fall of 7 feet in three hours. The fall in the tide commenced immediately after the center passed.

Key West, Florida was near the center of the tropical cyclone, October 10-12, 1909. The lowest barometer was recorded at 11:40 a. m. on the 11th. The sustained wind velocity was 74 miles for the hour ending 11:00 a. m., 66 miles per hour for the hour ending at 12 noon, (the hour in which the center of the cyclone passed the station) 63 miles for the hour ending at 1:00 p. m., and 29 miles for the hour ending 2:00 p. m. The wind continued to decrease in velocity, and was 12 to 13 miles per hour after 3:00 p. m.

Houston, Texas, was 15 miles to the right of the line followed by the center of the tropical cyclone, August 15-18,

1915. The center passed Houston, 5:25 a. m. August 17th, with the lowest barometer 28.20 inches. The sustained wind velocity for the hour ending 5:00 a. m. was 74 miles, for the hour ending at 6:00 a. m., during the passage of the center of the cyclone, was 80 miles, for the hour ending at 7:00 a. m. was 64 miles, for the hour ending 8:00 a. m. was 40 miles and it continued in the neighborhood of 40 miles per hour for several hours while the cyclone was curving to the right and bringing Houston into the right rear quadrant of the cyclone.

New Orleans, Louisiana, was on the fringe of the calm center in the tropical cyclone of September 29-30, 1915. The center of the cyclone passed at 5:50 p. m. September, 29th. The sustained wind velocity for the hours ending at 5:00 p. m. and 6:00 p. m. were each 60 miles per hour, for the hour ending at 7:00 p. m. was 41 miles per hour, and for the hour ending at 8:00 p. m. was 33 miles per hour. There were no high winds after the passage of the center of the cyclone.

Tampa, Florida was 15 miles to the right of the line followed by the center of the cyclone, October 23-26, 1921 with the center passing that station about 3:00 p. m. The sustained wind velocity was 53 miles per hour at 3:00 p. m., 37 miles per hour at 4:00 p. m., 43 miles per hour at 5:00 p. m., 41 miles at 6:00 p. m. and after that the velocities diminished rapidly.

New Orleans, La., September 19, 1947: This cyclone confirmed important features noted in the preceding cyclones. The public was warned over the radio to prepare for hurricane winds after the passage of the calm area. This advice came over my radio. During several days following the storm many persons asked me why the winds in this cyclone did not come up with hurricane velocities after the passage of the calm center. I referred them to the records of preceding cyclones in all of which the wind velocities in the rear of or near the calm center were not any thing like as great as the winds preceding the calm center.

The calm area in this cyclone passed over me at 29 Farnham Place, New Orleans, La. The calm or only light winds commenced at 10:00 a. m. and the calm lasted for about one hour. During the passage of the calm I scanned the sky closely; I did not see any blue sky nor any thing like the so-called "eye of the cyclone" which is supposed to be seen during the calm. There was nothing but clouds churning around in confused masses with lighter patches showing up occasionally. Rain had stopped falling and as a result it was lighter than during the heavy rain and high winds preceding the calm area. After the passage of the calm the winds came up with about half the force they had before the calm set in. No rain worthy of mention fell after the passage of the calm area. The false representation to the effect that hurricane winds follow the calm center probably results from the reports of persons 40 to 60 miles to the right of the path followed by the center of the cyclone; here it was noticed that the winds after the occurrence of the lowest barometer had higher velocities than the winds before the lowest barometer was recorded. Or they might have passed through the calm center and their station was carried quickly into the right rear quadrant of the cyclone by a sharp curve to the right in the path of the cyclone.

This is the second time that the center of a tropical cyclone has passed over New Orleans. The other was the tropical cyclone of September 29, 1915; the calm center of this cyclone passed over Tulane University, where the calm commenced with the barometer 28.10 inches at 5:30 p. m. and remained stationary until 6:00 p m., September 29, 1915. Four observers in the immediate vicinity of Tulane University reported almost a dead calm during that period. No one reported seeing a patch of blue sky.

Here we have two cyclones with calm centers in which no blue sky was seen. Heavy rain stopped with the commencement of the calm and there was little or no rain during or after the calm. No high winds occurred in the rear of the calm center. Clouds churned around in confused masses, occasionally breaking into lighter patches. The high winds of the right rear quadrant converging with the high winds in the right front quadrant ascend rapidly; the winds move around the calm center into the left part of the cyclone where they move on around the calm center and merge with the winds of the right rear quadrant. These movements of the winds around the cyclonic center would cause much confusion and breaks in the clouds over the calm area and as a result of this blue sky might at times be seen.

The so-called "eye of the storm" may be observed over the calm area at the earth's surface under certain conditions which are not always present. The lowest barometer at the earth's surface is in the calm area; the calm center is some 60 to 80 miles in the rear of the center of action in the cloud region where the energy of the cyclone is released; this is shown by the heaviest precipitation being recorded in that part of the cyclone. The core of the cyclone is inclined from the calm area at the earth's surface forward across the winds into the cloud regions over the area of greatest precipitation. This agrees with deductions made by Sir Napier Shaw (101). Since the calm area at the earth's surface is some 60 to 80 miles in the rear of the center of action in the cloud region which is over the heavy rain area of the cyclone, clear sky may be seen under some conditions over the calm area for the clouds in the rear half of the cyclone are breaking up at this time.

In all cyclones which moved forward without much curvature in their direction of travel the wind velocity diminished rapidly after the passage of the calm center not only directly behind the calm area but 20 miles to the right of the line followed by the center of the cyclone. In cyclones which curve sharply to the right so as to carry an observing station or a ship at sea, (which had passed out through the calm center of the cyclone) into the right rear quadrant some 40 or 50 miles to the right of the path then being followed by the center of the cyclone, hurricane winds may be encountered, sometimes with greater velocities than those which preceeded the calm area. In such cases the observers making the reports are not aware of the sharp curve made in the path of the cyclone, and not realizing their changed position as related to the center of the cyclone, report hurricane winds as having been encountered in the rear of the calm center; whereas, as a matter of fact, the winds they encountered were considerable distance to the right of the line flolowed by the calm center and they were in the worst part of the cyclone, the right rear quadrant.

CAUSES OF AND DISTRIBUTION OF PRECIPITATION

In describing the movement of the winds around the cyclonic center we have pointed out where the convergence oi the winds takes place and the area where convection of a violent nature is in operation. The importance of determining the points of local convection as related to the cyclonic center, and the position of the heaviest precipitation was emphasized by Sir Napier Shaw in 1922: He says: (100)

"We have no very satisfactory information about the distribution of precipitation in a tropical revolving storm. * * * Within the area of a cyclonic depression, convection may be of extreme violence, but it is not by any means uniform. The localizing of convection points apparently to the line along which additions to our knowledge might be made with advantage, and the accurate tracing of the position of rain areas must be of importance not only for the further investigation of the polar front but for many other meteorological investigations".

In our studies of tropical cyclones we have shown where convection takes place and the position of rain areas of different intensity in the cyclonic area as related to the cyclonic center (See Figure 7.) The discussion of the wind movement around the cyclonic center explains the causes of these important features.

Precipitation as it is occuring in tropical cyclones is not distributed around the cyclonic center as the rainfall shown along the path of the center on synoptic weather charts, after the passage of the cyclone might indicate, and as it is generally represented in contemporary works on meteorology. We find that very little precipitation occurs in the rear half of traveling cyclones. The greatest precipitation intensity in the larger cyclones is found 60 to 80 miles in front of the center and mostly to the right of the line along which the center of the cyclone was advancing at the time of its occurrence. In the smaller cyclones the greatest precipitation intensity is found not so far in advance of the center as in the larger cyclones, but it is also in front of the center and mostly to the right of the line along which the center of the cyclone was advancing. The precipitation distribution in cyclones moving through the Florida Straits is the same as that in the cyclones which traveled from the Gulf of Mexico into the interior. In all of the traveling cyclones the precipitation almost ceased with the occurrence of the lowest barometer and very little precipitation occurred in any part of the rear of the cyclonic area, whether traveling over water through the Florida Straits or moving inland on the Gulf Coast.

In the three cyclones which ceased to advance on moving inland the greatest hourly precipitation which had first occurred in the right front quadrant while the cyclone was advancing shifted to the rear of the cyclone as it stopped traveling and the precipitation does not show systematic distribution such as is found in traveling cyclones.

The causes for the precipitation occurring mainly in the front half and very little in the rear half of traveling cyclones can not be explained as the result of adjacent land and water areas. In all three of the cyclones which moved through the Florida Straits, and across the Florida Peninsula, where the winds in all the quadrants were almost continuously coming off of large bodies of water, we find the same general distribution of precipitation as related to the cyclonic center as we find in the other traveling cyclones moving over land areas. Furthermore, in cyclones which ceased to advance, the precipitation occurred in all parts of the cyclonic area and mostly in the rear half, notwithstanding all three of these cyclones occupied positions with regard to land and water areas the same as that occupied by traveling cyclones in which the precipitation was confined mostly to the right front quadrant and in which there was no precipitation in the rear half of the cyclone. The position of land and water areas as related to the direction in which the cyclone is traveling accounts for the intensity of precipitation in the cyclone, but not for the distribution of precipitation over the cyclonic area.

The temperatures in different parts of tropical cyclones traveling through this region do not vary greatly. These show that we must look elsewhere than to surface temperature differences for the causes and distribution of precipitation in traveling cyclones.

The mechanical and physical processes in these cyclones as shown by wind directions and their convergence and resultant ascension of air masses within the cyclonic area will account for the precipitation and its distribution and also for the greatest precipitation intensity occurring in the same position as related to the cyclonic center in all traveling cyclones.

We have already shown that the winds in the right rear quadrant of the cyclone blow, with high velocities, continuously during the life of the cyclone in the same general direction as that in which the cyclone is traveling. The winds in this quadrant certainly play an important part in the causes and distribution of precipitation in cyclones ,and the sources through which the precipitation is supplied and from which the supply of energy in the cyclone is derived. In the larger cyclones these winds constitute air streams at least five miles in depth, 250 to 400 miles in width, and 400 to 800 miles in length, extending in the line of advance of the cyclonic area. In this connection reference is made to Figure 7 where the air stream is represented by combining four of the larger traveling cyclones. The length of the air stream shown in the figure is 400 miles and it extends back some 400 miles additional as part of the air current in which the cyclone is traveling.

While the cyclone is traveling with a speed of 8 to 15 miles per hour the winds of the right rear quadrant of the cyclone come into the right front quadrant across the line dividing the cyclone into front and rear halves with sustained velocities near the earth's surface of 80 to 100 miles or more per hour extending back for some distance in the rear of the cyclone and velocities of 30 to 40 miles per hour or more often prevail for a distance of 400 miles or more in the right rear quadrant come into convergence with the winds of the right front quadrant as previously described. As a result of the convergence the moist air moving over the earth's surface in the winds coming through the right rear quadrant of the cyclone is forced to ascend in the right front quadrant of the cyclonic area and as it ascends is carried forward by its own momentum towards the front of the cyclone.

Sir Napier Shaw: In discussing the physical processes of weather refers to the momentum of air streams and says (62)

"It suggests that the great streams of air sometimes as much as 1,000 miles broad, are from their own momentum, features of greater importance from the dynamic point of view than the position of high and low pressure."

The momentum which the air streams in the right rear quadrant of the cyclonic area receive from the high velocities prevailing over a long stretch in that part of the cyclonic area, causes them, when they come into convergence with winds in the right front quadrant, to ascend through and mix with the winds which blow across the front of the cyclonic area. On account of the momentum, this air stream (See Figure 7) from the rear continues to move forward after the convergence, more rapidly than the cyclone is traveling, and it ascends and moves on at increasing elevation in the right front quadrant of the cyclonic area. This air stream, nearly saturated with moisture, expands and cools dynamically as it ascends, and this causes the condensation and precipitation of the water vapor and the liberation of latent heat, in the front and to the right of the line along which the center of the cyclone is advancing.

The latent heat released as the result of condensation retards the rate of cooling of the ascending saturated air, and therefore at the elevation where condensation is taking place the temperature is higher than it would have been had the cooling taken place at the adiabatic rate resulting from expansion not modified by the condensation of aqueous vapor, and the resulting release of latent heat. The ascending air mass from

CHARACTERISTICS OF TROPICAL CYCLONES

which precipitation is falling therefore has a higher temperature than the surrounding air at corresponding elevations and this causes the ascent and rarefaction of the air mass to become more rapid and further intensifies the condensation of moisture, and thus augments and intensifies the precipitation directly under the region where the air mass is ascending most rapidly. This takes place to the right of the line followed by the center of the cyclone and over an area directly in front of the line dividing the cyclone into front and rear halves extending sometimes more than 100 miles into the right front quadrant. (See Fig. 7).

From the foregoing the causes for the precipitation as found in cyclones are summed up as follows: The convergence of the winds from the right rear quadrant with those of the right front quadrant causes air to ascend; the momentum carries the ascending air mass forward into the front of the cyclonic area expanding and cooling as it reaches higher levels; the vapor of water in the nearly saturated air mass is rapidly condensed, giving excessive precipitation in the front and the right front quadrant as found in the eleven traveling cyclones studied. In this manner the enormous amount of energy required to produce the destructive forces in and keep the cyclone in action is supplied.

INFLUENCE OF ELEVATED LAND AREAS

The cyclone of August 15-22, 1915, moved into Texas from the Gulf of Mexico, and crossed the United States without any high pressure areas making their appearance to influence the precipitation in any part of the cyclone. We have in this instance a rare case where during a period of five days and covering a distance of nearly 2,000 miles, the precipitation in a tropical cyclone while traveling in the middle latitudes was influenced only by the cyclonic circulation, the carrying air current, and the topography of the country.

There were no anticyclonic or other weather conditions to influence the distribution of rainfall within the cyclone during its journey of some 2,000 miles, hence topography is the only outside factor besides the cyclonic circulation to be considered. The precipitation did not extend to any significant distance to the left of the line followed by the center of the cyclone, while on the right of that line we find precipitation occurring with the passage of the cyclone all the way to the Atlantic coast, the distance to the coast being 1,200 miles in some places. The rainfall along the right side of the path of the center of the cyclone showed very little change in the amounts from the time the cyclone moved inland until it reached the mountainous regions of western Arkansas, August 19th; here the circulation in the cyclonic area showed confusion and the intensity of the rainfall diminished to about half the amount previously recorded. When the cyclone reached the level lands of the central Mississippi Valley, August 21st, the rainfall intensity showed a marked increase, the amounts reaching 8 to 10 inches in Northeast Arkansas and Southeast Missouri, being nearly as great as when the cyclone moved in on the Gulf Coast at Galveston and Houston, Texas, August 17th. (See Figure 8). The heavy rainfall in the Central Mississippi Valley accompanied a decided increase in the intensity of the cyclone. The heavy precipitation was from water vapor which had been carried up from the Gulf of Mexico over the level valley lands by the air stream operating in the right rear quadrant of the cyclone. The winds forming the air stream had not met any obstruction to force their ascent until they converged with the winds of the right front quadrant over Northeast Arkansas and Southeast Missouri. This convergence of the winds caused the heavy precipitation and the increase in the intensity of the cyclone over that area. After the cyclone had passed over the Central Mississippi Valley and continued to travel northeastward it had the Appalachian Plateau and its mountains between it and the Gulf of Mexico and the Atlantic Ocean, its sources of water vapor. The cyclone diminished rapidly in intensity. Heavy precipitation occurred along the Appalachian Plateau; the moist air which was being drawn up from the Gulf of Mexico and the Atlantic Ocean was cooled in ascent over the mountains, and its moisture precipitated before reaching the front of the cyclone. The curtailment of the moisture supply of the cy-

CHARACTERISTICS OF TROPICAL CYCLONES 295

FIGURE 8—Rainfall during the passage of the tropical cyclone, August 17-22, 1915, across the United States. Broken line is path of center. Heavy rains shown to left of path in Oklahoma, Arkansas, Missouri, and Illinois, fell in the right front quadrant of the cyclone, August 19-21, 1915; see Table VIII, pages 89 to 96 and Figures 13 and 14, TROPICAL CYCLONES, Cline, The Macmillan Company, 1926. The cyclone curving to the east carried the center through the heavy rainfall.

clone was attended by a rapid and marked decrease in the intensity of the cyclone. As the left half of the cyclone moved over the Great Lakes there was no increase in precipitation on the left side of the path of the center of the cyclone. (See Figure 8.)

THE SAME CAUSE WILL ALWAYS PRODUCE THE SAME EFFECT

We have found from these studies that well defined phenomena repeatedly occur in the same portions of different cyclones with similar progressive movement in different years and in different localities whether moving in on the coast or over stations nearly surrounded by water. Some definite forces acting in the line of advance and within the cyclonic area itself cause well defined wind directions and heavy precipitation to occur in the same portions of all the traveling cyclones. Specific wind directions, high velocities, heavy precipitation, occur in the same parts of all traveling cyclones. It is a general maxim of physical science that the same causes will always produce the same effects. Therefore, the same distribution of the wind directions and wind velocities, and precipitation we have found in cyclones in this region will be found in cyclones which occur in other parts of the world.

ORIGIN AND DEVELOPMENT OF TROPICAL CYCLONES

The records we have show that cyclones moving through the Gulf and South Atlantic regions have nearly circular isobars, low barometer and a calm or little wind at the center, torrential rains and high wind velocities, the same as are found in cyclones traveling through the tropical regions, and the wind directions in the cyclonic area are similar to those found by Hall (67) in cyclones in the Caribbean Sea; and technically, until disproved if not true, it may be inferred that the conditions of pressure, wind, and precipitation found in these cyclones in the Gulf of Mexico and South Atlantic regions are in the main the same as those which prevail in cyclones in the tropics. From what we find in these cyclones we also infer that tropical cyclones do not differ in their development from the extra-tropical cyclones, except as modified by the influences of latitude and the differences in the moisture in the regions where the cyclonic development takes place.

Prof. Humphreys in discussing tropical hurricanes, says: (73)

"The physical causes of these storms, if they originate as seems probable, in the doldrums and between the counter-trades or similar winds, appears to be partly thermal and partly mechanical and their subsequent maintenance, after reaching the middle and higher latitudes, the same (largely mechanical) as that of any other cyclone of the same place."

We have in our study a series of cyclones the greater number of which, during the twenty-five years under consideration, have moved from the tropics and thence northward across the United States. When in tropical regions they are tropical in character, but in the middle latitudes they are extra-tropical, the same cyclone adapting itself to the region through which it is traveling. The principal differences between these cyclones while in the tropics and in the middle latitudes are differences in intensity which are determined partly by latitude affects and to a greater degree by differences in the amount of water vapor supplied.

The mechanical forces acting in these cyclones must necessarily be of the same nature both in the tropical and the middle latitudes, but modified by latitude effect, topography, and the position and intensity of high pressure areas in the neighborhood of the cyclone. In the middle latitudes the differences in the amount of water vapor supplied to the cyclone is determined to a great extent by the topography of the country between the cyclone and the source of moisture supply. Plateaus and mountains frequently cut off the feed line that supplies the moisture from which the energy of the cyclone is produced. (See Figure 8). An excellent illustration of this is found in the tropical cyclone which crossed the United States, August 17-22, 1915.

Differences in the temperature of the tropical regions and that of the higher latitudes, even over bodies of water, account to a great extent for the differences in the energy of the cyclone in the two regions. The mean temperature in the West Indies during the season of tropical cyclones is about 80 degrees, and at this temperature, a cubic foot of air can contain 11 grains of water. The mean temperature of the great storm tracks in the North Atlantic is about 55 degrees, and a cubic foot of air at this temperature can contain only 5 grains of water. A reduction in temperature amounting to 25 degrees in an ascending saturated air mass, from 80 degrees to 55 degrees in the tropics, would precipitate 6 grains of water from each cubic foot of air; while a 25 degree reduction in the temperature of an ascending saturated air mass from 55 degrees to 30 degrees in the North Atlantic would precipitate only 3 grains of water vapor from each cubic foot of air. With the same fall in temperature the amount of precipitation in the tropics would be twice that of the North Atlantic region. The difference in the temperature governing the amount of moisture supplied in the two regions combined with the latitude effect accounts for the observed difference in the intensity of the tropical cyclone in tropical and North Atlantic regions. There are instances in which a tropical cyclone shows as much intensity in the North Atlantic Ocean as when it was in the tropics. When moving northward over the Atlantic Ocean the air stream in the right rear quadrant keeps the cyclone supplied with warm air saturated with water vapor which furnishes the energy that keeps up the intensity of the cyclone. When the cyclone curves to the eastward the air stream in the right rear quadrant comes off land and cold water with the amount of water vapor in a cubic foot of air about half what it was in the air stream coming out of the warm regions of the south.

The development of tropical cyclones, it appears, may be brought about by large air currents running into the doldrums, the convergence of counter currents on the border of the trade winds, or air currents running in the same direction with different velocities which develop ascending air streams. The relative positions of the air currents are factors in determining the development and the progressive movement of the cyclones. These air currents may be originated by the differences in the thermal conditions of large masses of air or by

CHARACTERISTICS OF TROPICAL CYCLONES 299

an unequal distribution of pressure the seat of which may be far away from the region where the convergence of air streams takes place.

Algue (53), in his CYCLONES OF THE FAR EAST, has expressed the opinion that many of the cyclones in the China Sea are originated by opposing currents.

Doberck (55) in his work LAW OF STORMS OF THE EASTERN SEAS states that opposing currents develop tropical cyclones.

The air currents which converge and cause the air to ascend may be functioning in the upper air over the doldrums or on the border of the trade winds, and may not always extend to the earth's surface. Where we have the convergence of air currents and the resulting air stream cooling as it ascends, the liberation of latent heat, resulting from the continuous and excessive condensation and precipitation of water vapor, energetically keeps up the rarefaction of the air over the area of condensation and precipitation; the air moves in from all directions towards the area where convection is taking place, and when at a sufficient distance from the equator the deflective force of the earth's rotation acts on this inflowing air and we have the cyclonic development in action. The convergence of the winds in the cyclonic area then sets in at the earth's surface as found to exist in the cyclones charted in this study. (See Fig. 7).

Once the cyclonic movement is started the air current in which the cyclone is traveling (the trade winds in the tropics) will continue it in action after the primary cause has ceased to exist. This air current carries the newly developed cyclone forward brings the winds of the right rear quadrant into the line of advance of the cyclone and their convergence with the winds of the right front quadrant of the cyclonic area keeps up the ascending air in which the condensation and precipitation is taking place. This ascending air is carried forward by momentum into the right front quadrant of the cyclonic area and is attended by condensation and precipitation as we have found it in the same cyclones in the Gulf and South Atlantic regions. The cyclone when once started is continued in action in this manner by the action of the air current in which the cyclone is traveling. The course which the cyclone follows is determined by the trade winds over the Caribbean Sea, the summer monsoonal winds over the Gulf of Mexico and finally by the winds of the middle latitudes, except when there are abnormal movements of the upper air currents which overcome or break up the regular drift of the atmosphere for the time being, and cause the cyclone to follow abnormal courses, or die out as is shown where three cyclones ceased to advance on moving inland.

CONTINUOUS REDEVELOPMENT OF THE CYCLONE OVER THE AREA OF GREATEST PRECIPITATION

From the precipitation shown in the cyclones charted, we get some conception of the enormous amount of energy continuously being released and operating in the right front quadrant and directly in front of traveling cyclones. There is ample energy released there to account for the continuous redevelopment of the greatest intensity and to account for the observed phenomena attending such cyclones.

The energy supplied by the condensation and precipitation of moisture and release of latent heat in the ascending air in the right front quadrant of the cyclone does not of itself carry the storm forward, but it does determine the pressure minimum, the steepness of the barometric gradients, the wind velocities, and in fact the destructive forces of the cyclone. Furthermore, the position of the greatest precipitation intensity is indicative of the vigor of the carrying current; in traveling cyclones the greatest hourly precipitation intensity is well in the front of the line dividing the cyclone into front and rear halves. When cyclones are moving slowly or changing direction, the position of heaviest precipitation changes from one place to another corresponding to the changes taking place in the direction of the carrying air current. However, the position of heaviest precipitation is evidence of the carrying currents only when the cyclone is traveling over regions free from mountains which influence the position of heaviest rainfall. It appears from the evidence at hand that the depth and width of the air current in which the cyclone exists and which carries it forward has much to do with determining the extent of the cyclonic area.

The intensity of the cyclone depends on the position and amount of moisture supply as related to the center and direction of movement of the cyclone. The distance over which a cyclone will draw its moisture supply depends on the size of the cyclone and the length of the air stream in the right rear quadrant moving forward in the line of advance of the cyclone. The moisture supply may be curtailed, not only by distance from water areas, but when the air stream in the right rear quadrant, which carries the moisture, is drawn up over elevated regions causing the precipitation of the moisture before the air reaches the cyclonic area; or by changes in the direction of movement of the cyclonic area which causes the air stream in the right rear quadrant to come off land areas instead of water areas.

The area of lowest pressure at the earth's surface, in traveling tropical cyclones, occurs to the rear of the area over which the greatest condensation and precipitation takes place. There is no question but that the center of energy in the cyclone and the center of cyclonic action is continually renewed in the cloud strata over the region of greatest precipitation intensity and the occurrence of the lowest pressure at the earth's surface is to the rear of the center of action and release of energy in the cloud region.

The core of the cyclone is inclined from the low barometer at the earth's surface forward into the cloud region over the area of greatest precipitation as is indicated by the movement of the area of lowest pressure at the earth's surface towards the region of greatest precipitation intensity.

Shaw (101) in discussing convection in cyclones describes the manner in which rotating motion will extend downwards to the earth's surface and in conclusion says:

"Hence we may conclude that the core of a cyclone descending to the surface will not be along a vertical line but to a point on the surface displaced from the vertical across the wind of the lower layers".

The fact that precipitation almost ceases with the occurrence of the lowest barometer, and that there is only light precipitation in the rear of traveling cyclones, shows that there is very little ascending air in the region of or in the area of the lowest barometer which indicates the passing from the front to the rear of the cyclone.

The energy of the cyclone is released and comes into action over the area of greatest precipitation. The air stream in the right rear quadrant of the cyclone, which moves forward continually in the line of advance of the cyclone, is the mechanism which supplies the material from which the energy of the cyclone is produced. Through this air stream the traveling cyclone may draw its moisture supply from water areas as much as 600 to 1,000 miles to the rear of the cyclonic center.

When the tropical cyclone of August 17-22, 1915 was over the central Mississippi valley, in extra-tropical regions, torrential rains fell over the front and right front quadrant of the cyclone, in southeastern Missouri, northeastern Arkansas, and southern Illinois. (See Fig. 8). With these heavy rains the barometer fell rapidly and the cyclone increased greatly in intensity. The moisture supply from which the torrential rains were produced had been carried up over the comparatively level lands of the Mississippi Valley, from the Gulf of Mexico, more than 600 miles distant, by the air stream in the right rear quadrant of the cyclone. The cyclone at this time was much nearer to the Great Lakes than to the Gulf of Mexico, and notwithstanding the left half of the cyclone was approaching and traveled on directly over the Great Lakes there was no increase in the precipitation in the left half of the cyclonic area.

After the cyclone crossed the Mississippi Valley, the Appalachian Plateau came in between the cyclone and its source of water supply, the Gulf of Mexico and the Atlantic Ocean; the elevated regions encountered here caused the precipitation of part of the moisture being brought up from the Gulf of Mexico and the Atlantic Ocean before it reached the right front quadrant of the cyclonic area. Figure 8 shows the extent to which the Appalachian Plateau caused precipitation on its windward side as the moisture laden air stream from the water areas were drawn over that elevation to the right front quadrant of the cyclone. It is a fact worthy of note that as the left half of the cyclone approached and moved over the Great Lakes the precipitation diminished on the left side of the cyclone in the same proportion that it diminished on the right side of the cyclone.

Counter or opposing upper air currents sometimes neutralize each other and the current in which the cyclone is traveling fails; in such cases the area of precipitation extends over the cyclonic area and the cyclone fills up and dies out. Sometimes one air current completely overcomes the other, carries the ascending air stream from which precipitation is occurring in another direction; the cyclone redevelops and travels in the direction determined by the stronger air current. The backing of cyclones and loops in their paths are caused by changes in the direction of movements of the air currents at some distance above the earth's surface; air currents above the earth's surface determine the location of precipitation as related to the cyclonic center and the resultant direction in which the cyclone travels.

Sir Oliver Lodge (75) in referring to the energy of cyclones says:

"I do not find that people in general are aware of an important source of energy for the maintenance and intensification of cyclones, nor am I acquainted with a clear expression by a meteorologist that the condensation of aqueous vapor will suffice.

"Atmospheric pressure being a ton per square foot, the disappearance or collapse of a cubic foot of ordinary air would yield a foot-ton of work. The disappearance, by complete condensation of aqueous vapor in 760/12.7, say 60 cubic feet of atmosphere would yield about the same amount.

"If then the temperature of the saturated air fell from 18 to 12 C., by reason of condensation and rainfall, so that the vapor pressure diminished from 15.36 to 10.46 millimeters of mercury, a foot-ton would be generated in each 155 cubic feet of that region of atmosphere. Incidentally, the corresponding deposit of liquid would be 5 grains per cubic meter, or a rainfall of one-third inch from a vertical mile of air.

"Assuming that the above fall of temperature in the central region of a traveling cyclone is not excessive, the energy available in each cubic mile of it would be nearly a thousand million foot-tons."

The energy is not released in the central region of the traveling tropical cyclone as assumed by Lodge, but is released in the front and right front quadrant of these traveling cyclones.

The hourly precipitation in the larger cyclones, as shown on Fig. 7 exceeds one-third of an inch over an area 100 miles wide and 250 miles in length across the right front quadrant of the cyclone, 25,000 square miles. Half an inch to more than one inch of rain per hour falls over an area 50 miles wide and 150 miles long, 7,500 square miles, an area extending from the line followed by the center of the cyclone across the right front quadrant of the cyclonic area. The rain is continuous, in the right front quadrant, while the cyclone advances until the stations pass into the rear of the cyclone when the rain almost ceases. Rain commences 100 to 125 miles in front of the line dividing the cyclone into front and rear halves. The heaviest rainfall is recorded about 80 miles in front of and about 6 hours before the lowest barometer occurs which indicates that the stations have passed into the rear of the cyclone. The total amount of rain which falls during the passage of the front half of the cyclone ranges from 7 to 16 inches or more, at many stations, in 12 to 14 hours.

The area of lowest pressure at the earth's surface moves towards the locality where the precipitation intensity and convection is most marked, which as previously noted is the result of the forces which carry the cyclone forward. When the greatest precipitation intensity is far in advance of the line dividing the cyclone into front and rear halves the cyclone travels with greater speed than when the greatest precipitation intensity is nearer that line. When precipitation of any consequence is occurring in the rear half of the cyclone it travels slowly and if the greatest precipitation intensity occurs in the rear part of the cyclone it will die out or change its course.

HURRICANES, by Ivan Ray Tannehill (102) contains much valuable information on tropical cyclones, especially the history of such storms in the West Indies, the United States, and adjacent waters of the Atlantic Ocean, the Gulf of Mexico, and the Caribbean Sea.

STORM TIDES AND CURRENTS

The wind velocities in the left half of a cyclone are never strong as compared with the winds in the right half of the cyclonic area, and are not long sustained in the same direction over the same area. Furthermore, the progressive movement of the cyclonic area causes the winds over the greater part of the left half of the cyclone to recede continuously from the waves which these winds create; and thus their force for developing waves is diminished. Therefore there can be no great development of waves and swells in that part of a cyclone. In the right front quadrant of the cyclone the winds carry the waves across the front of the cyclone. Here the winds change directions so rapidly over the water as a result of the progressive movement of the cyclone that one direction does not persist over a distance of fetch of 75 miles during a period of as much as four hours. With this condition, even the high winds that occur in this quadrant, would not furnish sufficient energy for large wave development.

In the right rear quadrant of the cyclone, the winds blow in the main, in the same general direction as that in which the cyclone is traveling, at sustained velocities averaging 75 to 85 miles per hour for 18 hours and 55 to 60 miles per hour for 24 hours. It is in this part of the cyclonic area that the greatest length of fetch exists, for here the high winds persist mainly in the same general direction as that in which the cyclone is traveling, over a distance of 300 to 400 miles, and sometimes as much as 800 miles. The waves and swells are under the influence of the wind from the same general direction not only over this entire distance but also for another 100 to 200 miles, for waves and swells started in the rear of the right rear quadrant do not pass out of this quadrant for many hours during which time the fetch region has advanced a third to half a day's journey. For these waves and swells, then, the actual length of the fetch is 600 to 800 miles in large cyclones. When the size and speed of the cyclone and the velocity with which the waves and swells are moving forward are known, the length of these waves and swells may be computed roughly, from the following equation by Dr. Charles F. Brooks;

$$F = F + \tfrac{s}{w} F + (\tfrac{s}{w})^2 F + (\tfrac{s}{w})^3 F + \text{— — — —}$$

in which F is the length of fetch at any moment, S the forward speed of the cyclone and W the progressive movement of the waves.

This sustained direction and velocity of the wind in the right rear quadrant, prevailing during the life of the cyclone, furnishes the energy that develops and carries forward the larger waves of long length which move on through the smaller and shorter waves, pass on beyond the limits of the cyclone, and are carried by their inertia, in the line of advance of the cyclone, to shore long before even the front of the cyclonic area reaches there. At times as much as two days before dangerous conditions in the cyclone reach the coast.

It appears that no effort has been made to illustrate the differences in the relative sizes of the waves and swells produced by the winds which prevail in the different quadrants of the cyclonic area. Col. Reid as early as 1840, gave us a diagram illustrating the power of the winds of a cyclonic storm to give rise to swells moving outward in all directions from the center of the cyclonic area (97) but he did not differentiate between the size and persistency of the swells produced by the winds of the different quadrants of the cyclone. With a knowledge of the wind directions and velocities which prevail in the

CHARACTERISTICS OF TROPICAL CYCLONES 307

different quadrants of the cyclonic area we are able to illustrate the relative sizes and persistency of the swells which move out from the different parts of the cyclonic area, and these are illustrated in Fig. 9 where the size of the swells going out from each quadrant of the cyclone is shown to have a certain and well defined relation to the producing winds.

Powerful currents are developed in the right half of the tropical cyclone. Gas and whistling buoys with their chains and anchors included, weighing more than 30,000 pounds, located in the right half of the cyclone were carried varied distances ranging up to 8 to 10 miles. A gas and whistling buoy located a short distance to the left of the line followed by the center of the cyclone was not moved but its lights were extinguished. There were two tropical cyclones in which heavy objects were moved by swells or currents: That of August 13-17, 1915, and that of September 6-14, 1919. This feature in these cyclones is discussed under BAROMETRIC PRESSURE IN TROPICAL CYCLONES (storm tides and currents) pages 266-272 in this book. There were no tropical cyclones with movable objects in their right half except those mentioned above, during the 25 years under consideration. Some of the gas and whistling buoys mentioned above were in the left half of tropical cyclones at other times but were not disturbed or moved.

The large swells developed by the force of the continuous high winds, blowing in the same general direction as that in which the cyclone is traveling, run on to the shore far in advance of the cyclone, and evidently generate powerful currents in the right half of the cyclone. These swells and currents build up a storm tide on the coast, in the right front of the cyclone, commencing long before the cyclone reaches the coast line, and are thus one of the best guides for the forecaster to use in advising where the center of the cyclone will move inland and where the greatest danger to life and property will be encountered. As the cyclone moves inland the swells and currents carry the storm tide a considerable distance inland, depending on the terrain, sometimes more than 100 miles, building up a flood 8 to 15 feet deep to the right of the line followed by the center of the cyclone. Storm tides amounting to as much as 6 feet

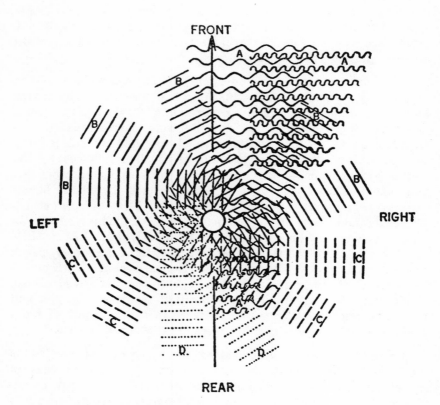

FIGURE 9—Relative sizes and direction of travel of waves and swells developed by the winds in tropical cyclones.

A. Swells of greatest length and magnitude sent forward by the winds of the rear right-hand quadrant and reach shore long before the cyclone reaches the coast line.

B. Swells and waves of moderate length and magnitude moving out to the right and left of the line of advance of the cyclone.

C. Swells and waves of smaller length and lesser magnitude in the rear segment of the cyclone.

This is Figure 9, Appendix, TROPICAL CYCLONES, Cline, The Macmillan Company, 1926.

have been recorded 200 miles to the right of the line followed by the center of the cyclone. (See Fig. 5). Storm tides do not occur to the left of the line followed by the center of the cyclone where the terrain is level. People should be moved out of low lands subject to overflow by the storm tides before the tide reaches dangerous proportions; I know from personal experience that this can be accomplished successfully.

Persons living in strong buildings in the area which will be traversed by the right half of the cyclone should be advised to close doors and windows tightly to prevent ascending air in the right front quadrant of the cyclone from lifting off the roofs and also to strengthen the building by making it more compact. The action taken to warn and save lives and property during the September 29-30, 1915 tropical cyclone, pages 158-159 this book, furnishes a good outline for the guidance of forecasters and others in their action in giving advice as how to save life and property in such emergencies.

GULF CYCLONE, SEPTEMBER 4, 1949

The Gulf storm of September 4, 1949, is of interest in this connection. The following extracts from press reports describing the damage along the Gulf coast as a result of the storm tide indicate that the storm was of considerable intensity:

Grand Isle, La.: The storm tide was three feet deep over the beaches and cottages and camps were damaged by winds and high tide.

Bay Saint Louis, Miss.; September 4th: High tides driven by the storm which buffeted the Gulf coast Sunday morning washed over the Bay Saint Louis-Henderson point railroad bridge blocking railroad service across the bay for several hours. Three miles of the Bay Saint Louis-Waveland highway was under water for a time during the storm.

Pascagoula, Miss., September 4th: Marshes and all low places were covered with water Sunday morning. Between here and Gautier at 7:00 a. m. 20 or more inches of water were standing on U. S. Highway 90, and water was also reported over low places in roads between here and Escatawpa. The water began to recede about noon.

Mobile, Ala., September 4th: Water piled up several feet deep on the Bay Bridge causeway skirting Mobile Bay. The causeway was reopened shortly before dark, after state convicts were brought in to clear the span of debris.

This was referred to in the newspapers as a "surprise storm". I can not understand why it was described as such. Storm tides reported above show that this storm was not different from other cyclones which moved in from the gulf. A graph comparing the storm tide with the regular predicted tides made by the U. S. Coast and Geodetic Survey for the Gulf coast, showing the storm and regular tides on the graph every three hours, should have given evidence that the storm was approaching the coast several hours before the storm reached the coast. The wind tide which was prevailing along that part of the coast would not have shown a continued steady rise in the tide but a storm tide would have shown a progressive rise. The storm tide would have shown up as a rise only to the right of the point on the coast where the center of the storm would move inland, and this soon would have shown whether the rise in the tide was from southeasterly winds or a storm in the Gulf.

When forecasting the weather all available material that will help serve the needs of the public should be utilized. Storm tides have always told me a true story. I found that others would not use the storm tides when I was using them successfully. Their use requires study and judgment. This will be recognized when the student reads understandingly my study on this subject: RELATION OF CHANGES IN STORM TIDES ON THE COAST OF THE GULF OF MEXICO TO THE CENTER AND MOVEMENT OF HURRICANES: U. S. MONTHLY WEATHER REVIEW, March 1920: 48: 127-142. When the reader understands the use of storm tides in warning the public where danger will be encountered, he will recognize that there is nothing else that will tell as accurately where the center of the cyclone will move in on the coast and where the death dealing and destructive storm tide will be most

CHARACTERISTICS OF TROPICAL CYCLONES 311

dangerous. The storm tide as it moves in on the coast is quick to show changes in the direction of movement of the cyclone. If the U. S. Monthly Weather Review cited above is not available the paper on storm tides can be found in Appendix, TROPICAL CYCLONES, Cline, 1926.

In this connection it is suggested that the student read the chapters on two cyclones: that of September 9-14, 1919, pages 170-174, this book; and that of June 21-22, 1921, pages 175-177 this book.

THE PROBLEM OF SO-CALLED "HURRICANE BUSTERS"

Newspaper dispatches, dated Miami, Florida, September 10, 1947, and on a few succeeding dates stated that plans were being made to break up a tropical cyclone (hurricane). We gathered from the press reports that the plan contemplated the dropping of dry ice or other cooling agencies into the center (eye of the cyclone) and cause record breaking rains and thus dissipate the cyclone. Our researches show that heavy rains could not have been produced in the center of the cyclone because the water vapor which supplies the energy that keeps the cyclone in operation does not pass into the center of the cyclone.

It appears that many meteorologists continue to hold to the false theory that the heaviest precipitation falls in and about the center of the cyclone. Synoptic weather charts representing conditions after the cyclone has passed shows heavy rainfall around the center of lowest barometer, but these do not give a true picture of what had taken place as the cyclone traveled forward. The heaviest rain actually falls directly in front of and to the right of the projected line along which the center of the cyclone was advancing. No rain worthy of mention falls in the calm center of the cyclone nor in its rear. In the preceding pages we have described the conditions as they actually occur in the different parts of the cyclone. (See Figure 7).

From newspaper accounts we infer that the "hurricane busters" did not have a correct knowledge of the area in the cyclone where the rain falls and the energy which keeps the cyclone in action is released. Press reports indicate a lack of understanding of the mechanical and physical forces operating in the cyclone. The rainfall shows the area on the ground over which the energy of the cyclone is released: Half an inch to more than one inch of rain falls hourly over an area of 25,000 square miles in the right front quadrant; this rain continues to fall for as long as 12 to 14 hours at many stations, with total amounts of 10 to 16 inches and sometimes more. The condensation of the aqueous vapor into rain releases the energy that keeps the cyclone in operation. This water vapor is supplied to the cyclone by an air stream some 230 miles in width and 400 miles or more in length continuously running like a river, with a speed of 40 to 100 miles per hour, into the right front quadrant of the cyclone (see Figure 7). The water vapor is fed into he cyclone by this air stream somewhat like the stream of water is fed into a turbine water wheel. This air stream feeding the cyclone with a speed of 40 to 100 miles per hour over a front of some 250 miles must be shut off before the cyclone can be dissipated. The operation of mechanical and physical forces in nature sometimes suddenly stop and break up a cyclone: In fact all cyclones are terminated by natural forces; by the movement of air masses that are so vast that man can not cope with them.

Observers in air planes making flights over the tops of tropical cyclones may learn something of their character and movement. What an observer in an air plane traveling 100 miles per hour or more can learn accurately remains to be established. He can determine the size of the cyclone and its height. He may learn something from the confusion of the wind tossed clouds. We do not believe that at the present time very accurate observations concerning conditions in a tropical cyclone can be made by an air plane flying over the top of the cyclone.

Radar? During the 25 years 1900 to 1924 inclusive there was little or no rain at any station in a traveling cyclone after the occurrence of the lowest barometer. No rain worthy of mention fell in the center of or in the rear of the center in any of

CHARACTERISTICS OF TROPICAL CYCLONES 313

the cyclones that continued to advance. There is no density of rainfall to identify the center of the cyclone.

Rain falls mainly in the right front quadrant and in front of the center, neither in nor around the center of the cyclone. The rain commences 100 to 125 miles in front of the center of the cyclone. Rain continues for 12 to 14 hours with 10 to 16 inches at some stations in the right front quadrant of the cyclone. Rainfall is heaviest to the right of the projected line along which the center of the cyclone is advancing, some 60 to 80 miles in space and 6 hours in time before the arrival of the cyclonic center or the occurrence of the lowest barometer at any station. In the larger tropical cyclones the hourly rainfall exceeds one-third of an inch over an area of 25,000 square miles in the right front quadrant; in the center of this area half an inch to more than one inch falls in one hour over an area 50 miles wide and 150 miles long extending across the right front quadrant, 7,500 square miles.

Some meteorologists appear not to be aware of or they question the above facts. They are referred to tables with brief discussions on pages 36 to 174 my book TROPICAL CYCLONES. In these tables all the weather data are printed exactly as they were recorded for each consecutive hour at every station that was in a tropical cyclone on the Mexican Gulf and South Atlantic coasts 1900 to 1924 inclusive. These tables show that rain almost ceased with the occurrence of the lowest barometer. The wind velocity diminished not only in the rear of the calm center but 20 miles to the right of the calm center after the occurrence of the lowest barometer. An important fact is that 40 to 50 miles to the right of the line followed by the center of the cyclone the wind velocities were as great or greater after the occurrence of the lowest barometer than before and the highest winds in the cyclone have been recorded in this area 6 to 8 hours after the occurrence of the lowest barometer at those stations.

"Men Against the Hurricane" in the October, 1950, issue of THE NATIONAL GEOGRAPHIC MAGAZINE is an interesting article, but some statements therein do not agree with observed facts about tropical cyclones. An illustration on page

545 purporting to show the movements of the winds around the center of the cyclone does not conform in any particular with the actual movements of the winds in a tropical cyclone. That conception of the movement of the winds around the low pressure was shown to be incorrect 100 years ago.

The statement "Spiraling Hurricane Winds Blow Hardest Near the 'Eye', Where the Pressure is lowest" does not agree with the location of recorded wind velocities in tropical cyclones. The highest wind velocities have been recorded in the right rear quadrant 40 to 50 miles to the right of the path followed by the center of the cyclone. Wind velocities have been much less after the occurrence of the lowest barometer directly in the rear of the calm center (Eye) and 20 to 26 miles to the right of the path followed by the center of the cyclone than before the occurrence of the lowest barometer.

On page 557 of the Magazine we find the following in connection with radar and the rainfall. "But the radar man says 'See these glowing crescents? They show where rain is heaviest; their curving shape reveals the hurricane eye.'

"You learn that radar pulses bounce back from water droplets in the air. Echo pulses shape up on the radar screen into patterns that trained observers can instantly identify.

" 'The center of a tropical cyclone shows up so well on radar,' Captain Ellsaesser explained, 'because the big and copious raindrops in the clouds around the eye toss back the strongest echoes.' "

The heaviest rain in tropical cyclones does not fall around the lowest barometer (Eye), the center of the cyclone. The radar man will frequently encounter heavy rains, one inch or more per hour, in separate localities in the front right hand quadrant of a traveling tropical cyclone but not around the center of low pressure.

We have made special detailed studies of all tropical cyclones that have moved in over the Mexican Gulf and South Atlantic regions during the 25 years 1900 to 1924 inclusive. There were eight traveling tropical cyclones with complete records and in all of these the heaviest hourly rainfall, in the larger cyclones was mostly about 6 hours in time and 60 to 80 miles in space in front of the lowest barometer; in the smaller traveling cyclones the positions of the heaviest hourly rainfall were relatively the same as in the larger cyclones.

There were three cyclones in which the heaviest rain shifted from the front to all parts of the cyclones and the lowest barometer shifted to the position of the heaviest rainfall. These cyclones when they stopped traveling had more of the theoretical movement of the winds and distribution of rainfall than when the cyclones were traveling. When cyclones are traveling the rainfall and wind directions are the same result of a combination of translation with rotation.

To substantiate what I have said in the foregoing regarding wind directions and velocities in traveling cyclones, reference is made to Figures 42 and 44, and the hourly rainfall to Figures 43 and 45, TROPICAL CYCLONES, Cline, THE MACMILLAN Company, 1926. In Figures 46 and 47 the reader wlil find the hourly wind velocities and directions and the hourly rainfall in tropical cyclones that stopped traveling and died out.

Those who are not satisfied with what the illustrations show are referred to the tables in the publication mentioned above wherein the weather elements recorded automatically are given for each hour just as they were recorded, while the station was in the cyclone, for every tropical cyclone that passed over the Mexican Gulf and South Atlantic regions in 25 years. The wind directions at the hour, the wind velocities for the hour, and the hourly rainfall are shown in their correct location in relation to the center of the cyclone on Figure 6, CHARACTERISTICS OF TROPICAL CYCLONES,

The data in the tables, which can not be questioned, show that there is little or no rain in the center nor rear of a traveling cyclone; neither is there rain of consequence in any part of the cyclone after the occurrence of the lowest barometer.

For the benefit of any who may not have access to my book TROPICAL CYCLONES I am giving here the positions of the heavy rains in their relations to the center of traveling tropical cyclones.

Tropical cyclone, September 26-28, 1906. New Orleans, La., 65 miles to the left of the path of the center, the heaviest rainfall was five hours in advance of the passage of the center. Mobile, Alabama, 50 miles to the right of the path of the center, hourly rains of about half an inch commenced about five hours in advance of the center, 1.34 inches fall one hour before the lowest barometer followed by half an inch for two hours after which the rain diminished rapidly.

Tropical cyclone, October 17-19, 1906. Key West, Florida, 50 miles to the left of the path of the center, hourly rainfall was moderately heavy commencing five hours in advance of the center and there was no rain worthy of mention after the occurrence of the lowest barometer. Jupiter, Florida, 30 miles to the left of the path of the center, moderately heavy rains commenced three hours in advance of the center with 1.02 inches one hour before the center, diminishing to .62 of an inch with the passage of the center.

Tropical Cyclone, September 19-21, 1909. New Orleans, La., 65 miles to the right of the path of the center, hourly rainfall was 1.00 inch per hour six hours in advance of the center, then less than one-fourth inch per hour with .07 of an inch during the passage of the center.

Tropical Cyclone, October 10-12, 1909. Key West, Florida, just a few miles to the right of the path of the center, moderate rains commenced six hours in advance of the center, with 2.82 inches two hours before, 1.32 inches one hour before and .62 inches in the hour during the passage of the center with no rain worthy of mention after that. This cyclone was curving sharply when passing Key West.

Tropical Cyclone, August 15-18, 1915. Houston, Texas, 50 miles to the right of the path of the center of the cyclone. Rain commenced getting heavy 8 hours in advance of the center when it was .67 inch then .76 inch and 6 hours in advance of the center it was 1.11 inches, 5 hours before .85 inch, then 2 hours before

.80 inch with .76 inch one hour before the center and very little after that.

Tropical Cyclone September 29-30, 1915. New Orleans, La., 14 miles to the right of the path of the center; the heaviest rain was six hours in advance of the center when the hourly rainfall was 1.59 inches, five hours before .84 inch, four hours before .91 inch, three hours before .30 inch, two hours before .38 inch, one hour before 1.05 inches, .60 inch during the passage of the center and none after that.

Tropical Cyclone, September 27-29, 1917. Pensacola, Florida, twelve miles to the left of the path of the center of the cyclone, a period of moderately heavy rains for four hours commenced 14 hours in advance of the center, then two hours before the center .75 inch one hour before, .88 inch during the passage of the center .58 inch then .80 and .40 and none after that. This storm curved sharply over Pensacola.

Tropical Cyclone, September 9-15, 1919. Key West, Florida, 20 miles to the right of the path of the center, moderately heavy rains commenced ten hours in advance of the center but rocerds were lost commencing one hour before the passage of the center.

Tropical Cyclone, June 21-23, 1921. The heaviest rain fell 16 hours before the passage of the center over Corpus Christi, Texas.

Tropical Cyclone, October 23-26, 1921. Jacksonville, Florida, 95 miles to left of the path of the center, heaviest rainfall was six hours in advance of the center and there was very little rain with the passage of the center. Tampa, Florida, 18 miles to the right of the path of the center, the heaviest rain fell eight hours in advance of the center and only light rain fell with the passage of the center.

APPENDIX A.

A CENTURY OF PROGRESS IN THE STUDY OF CYCLONES, HURRICANES, TYPHOONS.

By ISAAC MONROE CLINE
President's Address, American Meteorological Society
Pittsburg Meeting A. A. A. S. December 29, 1934.

Figures in Parenthesis refer to Bibliography

Notwithstanding the general importance of weather and weather changes in connection with all the activities of mankind, progress in the development of the science of Meteorology has been slow. The barometer was invented by Torricelli in 1643, but it was nearly 200 years later when the first important contributions were made to the study of cyclones and their individual characteristics. There were in the meantime contributions which were to have an important bearing on the study of storms and the movements of winds in cyclones.

The general circulation of the atmosphere plays such an important role in the development of the cyclone, its direction of travel, and the movements of the winds within the cyclone itself that Edmund Halley's contributions make him a pioneer in the development of Meteorology. He published, in 1686, studies that bear on winds in cyclones and also on dynamic meteorology. He was the first to attempt to explain the general circulation of the atmosphere as a result of the unequal distribution of the sun's heat over the surface of the earth. He studied the monsoons and contributed to the studies of the trade winds in which tropical cyclones have their origin. (1)

William Dampier was another pioneer. His book entitled "A Voyage Around the World," first edition, 1697, gives a valuable discourse on the winds, tides, and currents of the oceans. (2)

George Hadley brought attention, in 1735, to the deflection of the winds in their movements over the earth's surface as a result of the earth's rotation. (3) It appears that John Dalton, some 75 years later than Hadley, independently, and without knowledge of Hadley's contribution, made a clear explanation of the deflective effects of the earth's rotation. (4)

The first to draw a synoptic weather chart was Heinrich W. Brandes. He made a contribution to the laws of storms in 1820. (5) He pointed out that in a barometric depression the air moves towards the center of low pressure to establish equilibrium. He further discovered that cyclones travel from the west towards the east. (6)

It is now about 100 years since the publication of observed facts concerning the movements of winds in cyclones developed an intense interest in the study of this subject. The most prominent contributor was William C. Redfield. He was a man of extraordinary intelligence, vivid imagination, and possessed active and sound reasoning powers. In 1821 he traveled over a region where a severe storm had left the evidence showing the directions in which the winds had moved. The directions in which the prostrated trees were pointing were noted in his journal. He found the trees in the region near his home in one direction and those some distance from his home in the opposite direction. His analytical mind immediately conceived that there had been a vast whirlwind in the atmosphere and the observed facts bore out his opinion. He told his friends about his discovery, among whom was Prof. Olmstead of Yale University. Prof. Olmstead persuaded him to write a paper on the subject of storms. Redfield pointed out that the cyclone has a low pressure at the center and that low pressure areas are the cause of all great storms. (7) He was the first to define the cyclone as an extensive mass of air with a rapid movement counterclockwise around a calm center. Earlier writers had suggested that tropical cyclones are whirlwinds but Redfield established the fact by observation. Later he identified the regions in which tropical cyclones develop and called attention to their direction of travel. (8) Redfield has not received the credit his discovery and contributions in connection with the study of cyclones merit. This partly because he was slow in publishing his observations and conclusions and less important contributions to the subject appeared before the publication of his discoveries. He blazed the way for others to follow in the study of cyclones. (9) (10) He incorrectly assumed that in great storms the air moves in circles around the center. Among Redfield's last contributions may be mentioned "On the apparent necessity of revising the received systems of dynamical meteorology," (11) and "The study of a storm track of seven thousand miles." (12)

In 1828, Heinrich Wilhelm Dove published important contributions on cyclones. (13) He discussed tropical cyclones of the Southern Hemisphere as traveling systems with a clockwise movement of the winds, different from the counterclockwise movement of the winds in cyclones of the Northern Hemisphere. His celebrated book dealing with the laws of storms stood out as a monumental contribution to the subject. (14) This work was translated into English and went through several editions. (15) He considered the general circulation of the atmosphere as consisting of an equatorial current and a polar current and the weather encountered in the temperate zones as incidents in the conflict between these two currents. This is an early polar front point of view. (16)

James Pollard Espy commenced the publication of his contributions to meteorology about 1835. (17) He advanced the theory that every great atmospheric disturbance begins with the uprising of air as the result of being heated. (18 & 19) Then condensation sets in and resultant conditions intensitfy the ascending current; the air at the surface flows in beneath directly towards the center and rushes upward with increasing violence. He opposed the whirlwind theory advocated by Redfield, Dove, and others at the time. He contended that the motion of the wind was somewhat along the radii towards the center. He did not take into consideration the deflection of the winds as a result of the earth's rotation. He published his "Philosophy of Storms" in 1841, which contained much valuabel information. (20) In conjunction with other agencies he inaugurated a service of daily bulletins in which he showed the conditions of the weather in different localities. He urged the organization of a national weather service as early as 1841. Although Espy persisted in his error regarding the circulation of the winds in cyclones, his contributions contained much of real value.

Tropical cyclones in the South Indian Ocean were studied by Lieutenant-Commander William Reid who made valuable contributions to the subject. (21) He was the first to attempt to illustrate the movement of waves and swells sent out by the winds in the cyclone. He did not show any difference in the waves and swells sent out by the winds of the different parts of the cyclonic area. This chart was accepted as representing the correct movement of waves and swells sent out by the cyclone until very recently.

Studies of the cyclones of the Indian Ocean were continued by Captain H. Piddington. He coined the name "cyclone" now generally applied to storms of this character. The word is derived from a Greek word which means the "coil of a snake" and was selected to represent the combined circular and centripetal movements of the winds then thought to be characteristic of all systems of central low pressure. His findings were published in 1840 to 1876 and form a valuable contribution to the subject of cyclones. (22)

A contribution by G. Coriolis in 1835 discussed two important features: the relation of pressure to velocity under balanced forces and the deflective effect of the earth's rotation. (23)

In 1837 S. D. Poisson read before the Academy of Sciences, Paris, his important contribution, "On the Motion of Projectiles in the Air, Taking Into Consideration the Rotation of the Earth." (24)

Charles Tracy in 1843 contributed a paper in which the influence of the figure and revolution of the earth on the winds in cyclones was clearly set forth. (25)

Alexander Thom in 1845 made a contribution covering storms of the Indian Ocean south of the Equator in which he discussed their origin, causes, extent, rotary character, and other features. (26)

The appointment of Matthew F. Maury in 1844 to take charge of the National Observatory gave him the opportunity to collect a vast amount of valuable information on the winds and currents of the oceans and prosecute the studies of cyclones and meteorology in general. (27) His Physical Geography of the Sea, 1856, was translated into several languages.

W. Radcliffe Birt, as early as 1852, showed an unusually complete grasp of the movement of winds in cyclones. He was the first to point out that the direction of the winds in the cyclone would represent not simply the rotation but the combination of translation with rotation. (28)

Important contributions were made by Charles Meldrum of Mauritius. (29) He was the first to point out that the gyratory motion may be started by conflict of currents of air traveling in different directions. (30) Meldrum (31) made more than 50 contributions to the subject of cyclones. He was one of the first to attempt to show that the winds of a cyclone do not move in a circle about the storm center and in 1860 set forth that the winds incurve to the center.

Elias Loomis spent the years 1836-37 in Paris where he attended the lectures of Poisson and other noted scholars. About 1859 he published papers on meteorological subjects which had a bearing on cyclones. (32) His treatise on meteorology was the first American textbook on that subject. (33)

The study of meteorology was stepped up by the Alexander Buchan in his studies of the traveling cyclonic depression in its progress across the United States, the Atlantic Ocean, and Europe, March 13-22, 1859. (34) He was the author of an early textbook on meteorology. (35) The writer studied Buchan in college, 1881-82, while preparing to enter the weather service, Signal Corps, U. S. A.

Outstanding among contributors to the study of cyclones and meteorology in general in the 19th century is William Ferrel. As early as 1856 he contributed papers in which he discussed hurricanes (cyclones) and storms along advanced lines. (36) He advanced the view that adverse currents produce gyratory motions in the atmosphere at the place of their meeting. He says: "Hurricanes (tropical cyclones) then, and all ordinary storms must begin and gradually increase in violence by the action of some constantly acting force, and when this force subsides friction brings the atmosphere to a state of rest." He considers the release of energy resulting from condensation in the ascending air as the sustaining force in the cyclone and adds: "Therefore, as long as this ascending current can be supplied with air saturated with vapor, this continual rarefaction must take place, and also the ascent of the air in the middle of the hurricane and the consequent rush of air from all sides to supply its place:" In his later studies of cyclones he advanced the theory that the center of the cyclone is being continually formed in advance portions of the atmosphere. (37) At this time Ferrel espoused the incorrect theory that the air flowing in below from all sides ascends and flows out over the top of the cyclone. I mention this because many students accepted that view as correct and made studies of cyclones on that basis. He pointed out that neither the radial theory of Espy nor the strictly gyratory theory of Redfield, Reid, and others can be correct. (38) His theories regarding the movements of surface winds in cyclones have not been controverted. He describes the forces which determine the paths followed by cyclones. (39 & 40) In 1885 Ferrel made a remarkable contribution to advanced meteorology in which he dealt extensively with cyclones. (41) He shows clearly and conclusively the resultants of cyclonic and progressive motions on the winds in the cyclonic circulation. Ferrel used an illustration to show the combined effects of the general winds of the region on the winds in the cyclonic circulation which was probably the first diagramatic representation of the distortion of the winds in the cyclone resulting from the combination of translation with rotation. He pointed out that in front of a traveling cyclone the inclination of the winds towards the center would be decreased to a direction almost at right angles to the storm radius; while in the rear of the cyclone the resultant winds would be more radial, that is, blow more directly towards the center of the cyclone. (42) Although published half a century ago Ferrel's important contribution to meteorology has not received the recognition it merits.

Contemporary with Ferrel, students in Europe were working independently along similar lines. Among them C. M.

Guldberg and H. Mohn made important contributions to the study of cyclones, published 1876-1883. (43)

Dr. A. Sprung, in 1881, made a contribution dealing with particles moving freely over the earth's surface, especially their significance in meteorology. (44)

Severe droughts such as that of 1894 in the Great Plains Region are the result of cyclonic action. The high temperatures are the result of dynamic heating. We made a special study in 1894 of the action of cyclones in bringing about such droughts. (45) Since the publication in which this study appeared had a small circulation and no additional contributions to the subject have been made in 40 years we will give briefly the results of the study. Cyclones that move slowly southward over the Plains Region or remain nearly stationary, for three or four days in summer or sometimes longer, draw the north Pacific air across the Rocky Mountains to the low pressure area. In crossing the mountains the cooling of the air in ascent is attended by condensation and precipitation; the temperature of the ascending air mass is higher, at the elevation where condensation is taking place, than it would have been had no condensation occurred. The air reaches the eastern side of the Rocky Mountains very dry and with a relatively high temperature. We quote briefly from the Bulletin of the Philosophical Society, 1894, SUMMER HOT WINDS ON THE GREAT PLAINS, by Isaac Monroe Cline.

> Eight-tenths of the moisture of the atmosphere is below the crest of those mountains, and air in passing over them loses a large percentage of this moisture. This dry air in descending over the eastern slope, after having dissipated the cloud carried over, gains temperature dynamically nearly twice as rapidly, in a corresponding distance, as it cooled in ascending the western slope. In moving towards the low pressure area this dry air takes up the circulation around that area, is carried over the Plains Region from a northerly direction, and flows down over the eastern slope from a westerly and then a southerly direction depending on the trend of the isobars. The dry air is carried forward in the upper strata more rapidly than in the layers near the earth's surface and when thus carried out

over the moister and less dense air its tendency is to descend here and there through the air to the surface.

These currents (or air masses) in descending rapidly from great elevations gain a great deal of warmth and reach the earth with their initial dryness.

This study made 40 years ago carried a recommendation that forests be planted in the Great Plains for the purpose of reducing the geographical extent of these hot winds and their injurious effect on agricultural interests.

Contributions by Dr. Wm. von Bezold to the study of cyclones (46) and the distribution of atmospheric pressure, 1884-1906 (47) will aid the student materially in the study of cyclones.

The influence of mountain peaks 5,000 to 7,000 feet elevation in breaking up cyclones was studied in 1891 by W. L. Dallas. (48) The cyclone, after encountering a mountain, sometimes develops on the other side of the mountain one or two days later.

Extra-tropical cyclones crossing the Rocky Mountains have been studied by Professor Cleveland Abbe. (49) He found that the cyclone redevelops east of the mountains farther south than the original cyclone and adds: "The new cyclone is far less powerful because of its existence in the interior of a dry continent, and it will die away if not supplied by moisture."

Sir John Elliott published valuable contributions to the study of cyclones in the Bay of Bengal. (50) He estimated that cyclones of that region extend to a height of about 6,000 feet.

The origin and movements of cyclones in the West Indies by E. B. Garriott (51) contains much valuable material. He formulated rules regarding the movement of cyclones.

Sidney M. Ballou published in 1902 an exhaustive study of records of ship masters and others relative to the conditions observed in the center of tropical cyclones, such as the calm at the center and the eye of the storm. (52)

Studies of detailed observations at one station (Manila) are given by Rev. J. Algue, S. J., for six storms in his valuable book on cyclones of the Far East. He advanced the view that cyclones in the China Sea are originated by opposing currents. Algue says: (53)

> The wind directions will be less convergent on the front side, owing to the progressive movement of the cyclone, so that they will more approach a circular form than is the case in the rear. The directions of the winds on the two sides of the track undergo, owing to the direction of the track and the intensity of the progressive movement, changes which are all the greater, the greater the velocity of the cyclone.

Rev. B. Vines, S. J., in 1876 pointed out that the winds in the rear of tropical cyclones showed a great deviation towards the center and said: (54) "Not only at a distance from the vortex, but even in its vicinity."

W. Doberck (55) in discussing the storms in the eastern seas in 1904 points out the deviation of the winds from the spiral in the different parts of the cyclone and says:

> It also causes the wind to blow straight in towards the center behind the typhoon and to blow more across the path in front of the center .Less than half a mile up in the air the incurvature of the winds towards the center disappears in the average of the quadrants, but it still blows in towards the center in the rear. It is really the wind at this altitude that carries the typhoon along, for late in the autumn there are typhoons every year that move against the northeast monsoon, but we know that the monsoon is at times very shallow and there is a southwest wind above it. These typhoons sometimes disappear very suddenly; evidently when the northeast monsoon increases in depth and intensity.

H. Hildebrand Hildebrandsson reached the conclusion that the great currents of the atmosphere are indicated by the movement of cyclones and anticyclones and the mean directions of the clouds. (56) Hildebrandsson in conjunction with Teasserenc de Bort published extensive studies bearing on dynamic meteorology, important contributions (57) to the study of cyclones.

A monumental work by Dr. J. von Hann on climatology contains much valuable information bearing on the various features of the cyclone. (58).

Dr. Oliver Fassig contributed an instructive work on the paths and other features of tropical cyclones. (59) In an earlier publication he gives valuable information bearing on the forces that carry cyclones forward, and says: "The prevailing direction of the winds throughout the year in Puerto Rico is between east and southeast." (60)

Outstanding among contributors to the study of cyclones in recent years is Sir Napier Shaw. He has studied the life history of air currents, (61) air movements in cyclonic depressions, and other important features of the cyclone. (62) In his great work, Manual of Meteorology he has assembled an enormous amount of material dealing with the study of cyclones and their characteristics in various regions. (63) In a special contribution to the study of cyclones (64) he discusses among other things such important features as the maintenance of a cyclonic depression, the structure of the cyclone, the birth of a tropical cyclone, precipitation, death of a cyclone, height of the cyclone, and descending and ascending air. Shaw in commenting on a contribution by the late Lord Rayleigh (65) in which he set forth the conclusion that if fluid had any vorticity to start with and was carried along in a current of air the results would be the combination of translation with rotation, and says:

> I wrote Lord Rayleigh to explain that, although it might be exemplified in tornadoes and revolving tropical storms and in spite of the fact that tropical revolving storms sometimes passed continuously from their original condition to cyclonic depressions in our areas, I had been unable to find any evidence of the existence of revolving fluid, such as he described in the winds of a cyclonic depression of our latitude. He was unconvinced; and on further consideration I recognized that in looking for pictoral evidence of revolving fluid I had not taken proper account of the fact (which ought to have been obvious to any one during the last fifty years) that if the revolving fluid were carried along by a current of air the winds would represent, not simply the rotation, but the combination of translation with rotation.

We are at a loss to understand, with the findings published by Ferrel in 1885, Lord Rayleigh in 1917, as set forth by Shaw in 1922, and later substantiated by satistical studies of winds in cyclones, (69) why some of our meteorological publications continue to use charts representing an instantaneous uniform inward spiral movement of winds about the center in the cyclone. Such an inward uniform spiral movement does not take place in traveling cyclones.

Mrs. E. V. Newnham, of the Meteorological Office Air Ministry, London, (64) has published a valuable contribution to the study of tropical cyclones in the various parts of the world. She summarizes the studies of cyclones for different regions.

Major E. H. Bowie and R. H. Weightman, (66) contributed valuable and instructive information on the types of cyclones in the United States.

A brief but important contribution to the study of cyclones is that of Maxwell Hall. (67) In discussing the movements of winds in tropical cyclones at Jamaica, he says:

> But in most tropical hurricanes the incurvature is least in front and greatest in rear.

Studies of kite observations by Willis Ray Gregg (68) contain much valuable material for use in connection with studies of the depth of general air currents in which cyclones travel after passing into the United States. The observations of winds aloft published by Gregg show that as far north as Broken Arrow, Oklahoma, monsoonal winds prevail in the summer to a depth of two or three miles and at times the depth is still greater before the prevailing westerly drift of the atmosphere overcomes the drift of the monsoonal winds. (69) In his work on Aeronautical Meteorology he gives information on the winds of the upper air east of the Rocky Mountains which has an important bearing on the travel of cyclones. (70)

A study of the tides which build up on the coast as a result of the swells sent out by the winds in the cyclones in the Gulf of Mexico and adjacent regions by Isaac Monroe Cline, (69) (71) has brought out facts of great value in forecasting the intensity and movement of tropical cyclones and the areas which will be flooded by dangerous storm tides.

Rev. L. Froc has published the tracks of 620 tropical cyclones in the Far East. (72)

The contributions of Professor W. J. Humphreys in his works, Physics of the Air, (73) contain information of great value in the study of cyclones.

Dr. Charles F. Marvin has made contributions which emphasize certain important features of the cyclone. (74) Professor Marvin says:

> It is well known that Highs and Lows travel at considerable velocities in definite directions. Therefore the actual motion of the winds over the ground will be those compounded from (1) the motions appropriate to the system of curved isobars and (2) the motion of the systems as a whole.

In discussing the energy of cyclones Sir Oliver Lodge says:

> I do not find that people in general are aware of an important source of energy for the maintenance and intensification of cyclones, nor am I acquainted with a clear exposition by a meteorologist that the condensation of aqueous vapor will suffice.
>
> Atmospheric pressure being a ton per square foot, the disappearance or collapse of a cubic foot of ordinary air would yield a foot ton of work. The disappearance, by complete condensation, of the aqueous vapor in 760/12.7, say 60 cubic feet of atmosphere would yield about the same amount.
>
> If then, the temperature of the saturated air fell from 18° to 12° C., by reason of condensation and rainfall, so that the vapor pressure diminished from 15.36 to 10.46 millimeters of mercury, a foot ton would be generated in each 155 cubic feet of that region of atmosphere. Incidentally, the corresponding deposit of the liquid would be 5 grams per cubic meter, or a rainfall of one-third inch from a vertical mile of air. (69) (75)

Studies of cyclones in the South Pacific and Indian Oceans and in the Northeast Pacific between Hawaii and the coast of Mexico have been made by Stephen S. Visher. (76) He has also made a study of typhoons with charts of normal tracks. (76) Another publication by Visher contains a study of tropical cyclones of the Pacific Ocean. (77)

Cyclones of the West Indies have been studied in detail by Charles L. Mitchell. (78) He discusses the track, precursory signs, the average movement, the localities where they develop, forecasting, and other features of tropical cyclones. His work is a valuable contribution to the subject.

Studies, Tropical Cyclones, by Isaac Monroe Cline, in which the integration method was used for the first time in charting data in cyclones, was privately published by authority of the Secretary of Agriculture in 1926. (69) The results of these studies are summed up in a review of the book by Dr. E. W. Barlow, in the Journal of the Royal Meteorological Society, London. (This review is given in full in Chapter XXXVI in this book.)

Studies of tornadoes carried on by Col. John P. Finley over a period of 50 years form a valuable contribution to the study of cyclones. (80) (81)

In his studies of Aerography, Alexander McAdie gives material which will aid the student in the study of cyclones. (82)

To the name Bjerknes, father and son, must be credited important modern contributions to the study of the dynamics of cyclones, (86) (87) (88). They gave new and accurate expression to the thesis that the life history of cyclones involves the dynamic interplay of neighboring air masses of differing energy content due to diverse origins, and a fresh spirit and stimulus were thereby instilled in modern meteorology. The elements of the polar front theories are old, and many of the relations involved have been in use by forecasters for many years, but lacking the clear definitions and explicit terminology evolved by Bjerknes the usefulness of the ideas was limited and it was difficult if not impossible for students to exchange opinion fully and clearly. The meteorological world owes a debt of gratitude to the Norwegian school for the gift of a terminology and a formulary of basic notions which for the first time enables meteorologists to speak to each other in full exchange of ideas regarding the dynamics of cyclones.

Yosiki Hosiguti 1926 to 1928 (89) made contributions to the study of cyclones of great value to the student of this subject. He gives observations in several cyclones and makes deductions therefrom. In speaking of the energy of the cyclone he says:

> Therefore it may be said that the energy supply by means of the condensation of vapor is more than compensates the dissipation of energy, and no use to consider another source of energy besides rainfall due to converging current.

Studies of air currents, the movements of the upper air, and of the characteristics of air masses and their behavior add materially to our knowledge of cyclones. Studies of air currents and the movements of the upper air, bearing on cyclones, have been contributed by: Sir Napier Shaw (61) (62) (63); Willis Ray Gregg (68) (70); Isaac Monroe Cline (69); Oliver L. Fassiz (53); and valuable contributions are found in International Cloud Observations . (84) Valuable studies have been carried on at Blue Hill Observatory (85) under A. Lawrence Rotch, H. Helm Clayton, Alexander McAdie, and Dr. Charles F. Brooks. Important contributions to the study of air mass analysis have been made by: J. Bjerknes (86); V. Bjerknes (87); J. Bjerknes (88); H. C. Willet (90); C. G. Rossby (91); H. C. Willet (92); and F. W. Reichelderfer. (93)

Special studies of air currents, the movement of the upper air, and the characteristics of air masses and their behavior are being prosecuted not only by the United States Weather Bureau but by such educational institutions as the Massachusetts Institute of Technology, Cambridge, Mass., and the California Institute of Technology, Pasadena, Cal.

AIDS IN FORECASTING MOVEMENTS AND DESTRUCTIVE AGENCIES IN TROPICAL CYCLONES

Statistical facts brought out in our use of the integration method in assembling data recorded in tropical cyclones, (69) (71) enable us to summarize certain outstanding features concerning the clouds, winds, precipitation, and storm tides, for use in forecasting the action, movement, and intensity of tropical cyclones, viz:

1. Cirrus clouds well in advance of and during the passage of the larger traveling cyclones have the same general

directions as that in which the cyclones are traveling. The cirrus clouds moving with the cyclone indicate that the air currents or masses which carry the larger cyclones forward extend to the cirrus level. In the larger cyclones the lower clouds not more than half a mile above the earth's surface move around the center of the cyclone but are not inclined towards the center. The alto-cumulus and alto-stratus clouds recorded in the front of the larger cyclones are inclined away from the cyclonic center which indicates that the height of the circulation in these cyclones is between two and three miles. In the smaller cyclones the cirrus clouds follow their prevailing direction from west towards the east which is often across the path of the cyclone. This indicates that the air currents or air masses which carry the smaller cyclones forward do not extend to the cirrus level. In smaller cyclones there is no incurvature of the lower clouds towards the center in front of the cyclone; these clouds are not more than half a mile above the earth's surface. The height of the cyclonic circulation in the smaller traveling cyclones is less than two miles, often not more than a mile.

2. Winds in the rear right-hand quadrant of the cyclone have the same general direction as that in which the cyclone is traveling and continue so during the life of a travelling cyclone (See "Tropical Cyclones—Cline" page 218.) The intensity of the cyclone is determined by the length of the fetch, over water, of the winds in the rear right-hand quadrant of the cyclone; these winds carry the moisture to the front of the cyclone and thus supply the energy which keeps up the continuous redevelopment of the cyclone over the area where the greatest precipitation is taking place. When the length of the fetch over water, increases, the intensity of the cyclone will increase. On the contrary, when the length of the fetch, over water, is shortened by land areas the cyclone will diminish in intensity.

3. The precipitation results from the forced ascension of the air in the front part of the cyclone, caused by the convergence of the winds from the rear right-hand quadrant with the winds in the front of the cyclone. The rain area is found only in the front and front right-hand quadrant of cyclones traveling in straight lines and mainly in the front right-hand quad-

rant when the cyclone is curving to the right. The extent of the rain area indicates the size, and the amount of precipitation in a given time, hourly, the intensity of the cyclone. In the larger traveling cyclones the rain commences 200 to 300 miles in front of the center and the greatest intensity is as a rule about 100 miles in front of a line dividing the cyclone into front and rear halves. The rain almost ceases at any station when it passes into the rear of the cyclone. In the smaller traveling cyclones the greatest intensity of precipitation is about 40 miles in front of the line dividing the cyclone into front and rear halves. In a cyclone that is traveling rapidly the precipitation begins farther in advance of the center of the cyclone than when the cyclone is traveling slowly.

4. The center of the cyclone moves towards the area of heaviest rainfall. As long as the heaviest rainfall occurs in the front half of the cyclone, the cyclone will continue to travel and its center will pass over the area where the rainfall has been the greatest.

5. Mountain ranges half a mile or more in height interfere with the wind movements in cyclones and break up or disorganize the winds which form the air stream in the rear right hand quadrant (the feed line) and cause the cyclone to diminish in intensity. When the air current which carries the cyclone forward fails or is checked in its progress by an opposing current, the position of the convergence of the winds is thrown backward over the cyclonic area, the precipitation is then more uniformly distributed around the center of the cyclone than is found in traveling cyclones and the heaviest precipitation often occurs, in such cases, to the rear of the cyclonic center. Under such conditions the cyclone ceases to be a traveling cyclone, the center, that is, the lowest barometer, shifts to the location of the heaviest precipitation, backs, fills up, and dies out; except the cyclone is sometimes after one or two days picked up by another air current, then the cyclone becomes a traveling cyclone and usually moves off in another direction.

6. The winds, 200 to 300 miles in front of the center of the cyclone, have a direction which is at right angles to the projected line along which the cyclone center is traveling. No

incurvature of the winds towards the center has been noted at a station directly in front of a traveling cyclone during 12 to 20 hours preceding the arrival of the cyclone center at the station.

7. The greatest wind velocities are recorded some distance to the right of the line of advance of the center and to the rear of the line dividing the cyclone into front and rear halves. The velocities of the winds in the right hand part of a cyclone are always greater than those in the left hand part of the cyclone. Cyclones are carried forward by the general air currents of the latitudes in which they are traveling and these currents travel with speeds of 10 to 25 miles per hour or more. The air currents superimpose their rate of travel upon the velocities of the winds developed by the cyclonic circulation. For example, in a cyclone being carried forward in an air current traveling at the rate of 25 miles per hour the winds in the right hand part of the cyclone would have velocities, approximately 50 miles per hour greater than the winds in the left hand part of the cyclone. That is the carrying air current would increase the wind velocities something like 25 miles per hour in the right hand part of the cyclone and diminish the winds 25 miles per hour in the left hand part of the cyclone.

8. Winds of hurricane force have not been recorded, except near the immediate center, to the left of the line followed by the center of the cyclone.

9. Winds at stations passing out of the cyclone directly in the rear of its center diminish in velocity immediately after the passage of the cyclone center.

10. The storm tide which is built up in advance of the arrival of the cyclone is found only in the right-hand front and directly in front of the cyclone center. The tide is caused by the swells sent out and forward by the winds which prevail as continuous air streams in the rear right-hand quadrant of the cyclone. No storm tide is built up to the left of the line along which the cyclone is traveling at the time.

11. When the rise in the storm tide shifts position this indicates that the cyclone has changed direction of travel with its

center moving towards a point immediately at the left of the last reported rising tide.

12. The center of the cyclone moves in on the coast at the left edge of the area over which the storm tide continues rising. Cyclones which curve sharply to the right sometimes leave a relatively high, but rapidly falling tide to the left of the line followed by the center but no material damage results from such a tide.

13. Storm tides must be checked closely against the graph of the predicted tides. The storm tide is the excess of the tide over the predicted tide at the time. The regular daily tides of the coast region show up on the storm tides. During the approach of a cyclone a falling tide at the time of low tide must not be mistaken as a falling storm tide. At the time of low tide stations to the right of the projected line along which the center of the cyclone is traveling will often report a stationary or falling tide, but by comparison of the graph of the reported tide with that of the predicted tide it will be seen that the storm tide is steadily rising. On the contrary at stations to the left of the projected line along which the cyclone is traveling a rising tide may be reported at the time of high tide when by comparison with the predicted tides it will be found that the storm tide is falling.

14. Tropical cyclones are developed by the convergence of counter air currents on the border of the trade winds, by large air masses running into the doldrums, and by air currents traveling in the same direction with differing velocities, any of which cause the air to ascend and when a sufficient distance from the equator the deflective force of the earth's rotation brings the cyclone into action. The prevailing winds of the region come into play and we have the principle of the turbine operating to continue the cyclone in action. (See Figure 7). The broken lines represent the stream of moist air which flows up through the rear right hand quadrant of the cyclone. The existence of this air stream is established by automatically recorded wind directions shown by the short arrows in that part of the cyclone. This stream of moist air moving with a speed of 30 to 100 miles per hour converges upon the winds of the

right hand front quadrant of the cyclone and is thrown upward causing a rapid ascent of the air over that area. The air cools as its ascends, latent heat is liberated as the result of condensation of the moisture by cooling in the ascending air mass; this increases the rapidity of the ascending air mass and causes more rapid condensation of the moisture. In this way the air stream furnishes the power or energy that keeps the cyclone in action. As long as the air stream can supply the air saturated with moisture the redevelopment of the cyclone will continue over the region of greatest precipitation.

BIBLIOGRAPHY

1. An Historical Account of the Trade Winds and Monsoons. E. Halley. Philosophical Transactions, Vol. XXVI. pp. 153-168. London, 1680-1687.

2. A New Voyage Around the World. William Dampier. Fifth Edition corrected. London. 1703.

3. Concerning the Cause of the General Trade Winds. George Hadley. Philosophical Transactions. Vol. XXIX. 1735-1736. pp. 58-62. Also in Mechanics of the Earth's Atmosphere. Smithsonian Collections. Cleveland Abbe. Washington, D. C. 1910.

4. Meteorological Observations and Essays. John Dalton. Second Edition. Manchester, Eng. 1834.

5. Ueber die Entstehung des Regens und de Sturme. H. W. Brandes. 8 vo. Leipz. 1820.

6. Dissertatio physica de repentinis variationibus in pressione atmosphaerae observatis. H. W. Brandes. Leipz. 1826.

7. Prevailing Storms of the Atlantic Coast and the Northeastern States. William C. Redfield. American Journal of Science and Arts. Vol. XX. pp. 17-51. New Haven, Conn. 1831.

8. Hurricanes and Storms of the United States and West Indies. William C. Redfield. American Journal of Science and Arts. Vol. XXV, pp. 114-121 and Vol. XXVI. pp. 122-135. New Haven, Conn. 1834.

9. On the Whirlwind Characteristics of Certain Storms. William C. Redfield. Journal of the Franklin Institute. Vol. XIX. pp. 112-127. Philadelphia, 1837.

10. On Three, Several Hurricanes of the Atlantic, etc. William C. Redfield, New Haven, Conn. 1846.

11. On the Apparent Necessity of Revising the Received Systems of Dynamical Meteorology. William C. Redfield. Proceedings of the American Association. pp. 366-369. 1850.

12. A Storm Track of Several Thousand Miles. William C. Redfield. American Science. Vol. II. pp. 47-50. Cleveland, Ohio. 1854.

13. Ueber Barometriche Minima. Heinrich Wilhelm Dove. Pogg. Ann. Vol. XIII. p. 596. 1828.

14. Ueber das Gesetz der Sturme. Heinrich Wilhelm Dove. Theil I. 8 vo. Berlin. 1857.

15. Law of Storms Considered in Connection with the Ordinary Movements of the Atmosphere. Second Edition. Translated with the Author's Sanction (H. W. Dove) and Assistance by R. H. Scott. 8 vo. London. 1862.

16. Manual of Meteorology. Sir Napier Shaw. Vol. I. p. 290. Cambridge, Eng. 1926.

17. Contributions of James Pollard Espy on Storms. Journal of the Franklin Institute. Philadelphia. Vol. XVI. pp. 4-6. 1835; Vol. XVII. pp. 386-293. 1836; Vol. XVIII. pp. 100-108. 1836; Vol. XIX. pp. 17-21. 1837; Vol. XXIII. pp. 38-50, 149-158, 217-231, and 289-298. 1839.

18. Contributions on Storms. James Pollard Espy. Reprint by British Association, part 2. pp. 30-39. London. 1840.

19. Contributions on Storms. James Pollard Espy. American Journal of Science. Vol. XXXIX. pp. 120-132. 1840; Vol. XL. pp. 327-332. 1841.

20. The philosophy of Storms, James Pollard Espy. 8 vo. Boston, 1841.

21. The Progress of the Development of the Law of Storms and of the Variable Winds. Lieutenant-Commander William H. Reid. London. 1849.

22. Sailor's Horn Book for the Law of Storms. Captain H. Piddington. London and New York. 1848.

23. Sur la Maniere d'establir les Differens Principles de Mecanique pour des les Systems de Corps en les Considerant Come des Assemblages de Molecules. G. Coriolis. Jour. Ecole Polytechn. Vol. XV. pp. 93-125. Paris. 1835. And Sur les Equations du Mouvement Relatif des Systemes de Corps. G. Coriolis. Jour. Ecole Polytechn. Vol. XV. pp. 142-154. Paris. 1835.

24. Memoire sur le Mouvement des Projectiles dans l'air en Ayant Egard la Rotations de la Terre. S. D. Poisson. Jour. Ecole Polytechn. Vol. XVI. pp. 1-68. Paris. 1838. Also Mechanics of the Earth's Atmosphere. Smithsonian Collections. Cleveland Abbe. Washington, D. C. 1910.

25. On the Rotary Action of Storms. Charles Tracy. American Journal of Science and Arts. Vol. XLV. pp. 65-72. New Haven, Conn. 1843. Also Mechanics of the Earth's Atmosphere. Smithsonian Collections. Cleveland Abbe. Washington, D. C. 1910.

26. An Inquiry Into the Nature and Causes of Storms in the Indian Ocean South of the Equator; With a View of Discovering Their Origin, Extent, Rotary Character, Rate and Direction of Progression, etc. Alexander Thom. 8 vo. pp. 351. London. 1845.

27. Gales, Typhoons, and Tornadoes. Matthew F. Maury 4 to. Washington, D. C. 1851.

28. Hand Book of the Law of Storms. W. Radcliffe Birt. London, 1852.

29. Contributions to the Meteorology and Hydrography of the Indian Ocean. Charles Meldrum. Part I. 4 to. Mauritius. 1856.

30. On the Rotation of Wind Between Oppositely Directed Currents

of Air in the Southern Indian Ocean. Charles Meldrum. Proceedings of the Meteorological Society. London. 4 to. pp. 322-324. 1869.

31. Bibliography of Meteorology. Part 4. Storms. Signal Office. Washington, D. C. 1891.

32. On Certain Storms in Europe and America, December, 1836. Elias Loomis. Smithsonian Contributions to Knowledge. 4 to. ,Washington, D. C. 1859.

33. Treatise on Meteorology. Elias Loomis. New York. 1875.

34. Remarkable Storm of March 13-22, 1859. Alexander Buchan. Journal of the Scottish Meteorological Society. Edinburg. II. pp. 198-213. London. 1869.

35. Handy Book of Meteorology. A. Buchan. London. 1867.

36. An Essay on the Winds and the Currents of the Ocean. W. Ferrel. Nashville Journal of Medicine and Surgery. Vol. XII. Nos. 4 and 5. October and November, 1856. Nashville, Tennessee. Also Professional Papers of the Signal Service. No. XII. 1882. Washington, D. C.

37. The Motions of Fluids and Solids Relative to the Earth's Surface. W. Ferrell. American Journal of Science. Second Series. Vol. XII. pp. 27-51. May, 1861. Also Professional Papers of the Signal Service No. XII. Washington, D. C. 1882.

38. Cause of the Low Barometer in the Polar Regions and in the Central Parts of Cyclones. W. Ferrel. Nature. London. Vol. 4. pp. 226-228. 1871. Also Professional Papers of the Signal Service. No. XII. 1882. Washington, D. C.

39. Cyclones, Tornadoes, and Water Spouts. W. Ferrel. Coast and Geodetic Survey Report for 1878. Appendix. 10. American Journal of Science. Vol. XXII. July, 1881. Also Professional Papers of the Signal Service. No. XII. 1882. Washington, D. C.

40. Relation Between the Barometric Gradient and the Velocity of the Wind. W. Ferrel. American Journal of Science and Arts. Vol. VIII. November, 1847. Also Professional Papers of the Signal Service. No. XII. 1882. Washington, D. C.

41. Recent Advances in Meteorology. W. Ferrel. Annual Report of the Chief Signal Officer. Part II. 1885. Washington, D. C. 1886.

42. Popular Treatise on the Winds. W. Ferrel. New York. 1889 Second Edition. 1911.

43. Etudes sur les Movements de l'Atmosphere. C. M. Guldberg et H. Mohn. Christiana. 1876. Revised 1883-85. Also Mechanics of the Earth's Atmosphere. Smithsonian Collections. Cleveland Abbe. Washington, D. C. 1910.

44. On the Paths of Particles Moving Freely on the Rotating Surface of the Earth and their Significance in Meteorology. Dr. A. Sprung. Wiedemann's Annalen der Physik and Chemie. New Series Vol. XVI.

pp. 128-149. Leipz. Also Mechanics of the Earth's Atmosphere. Smithsonian Collections. Cleveland Abbe. Washington, D. C. 1910.

45. Summer Hot Winds on the Great Plains, U. S. A. Isaac Monroe Cline. Philosophical Society of Washington. Bulletin. Vol. XII. pp. 309-348. Plates 4-6. Washington, D. C. 1894.

46. On the Theory of Cyclones. Prof. D. Wm. von Bezold. Sitzungberichte of the Berlin Academy. pp. 1295-1317, 1890. Also Mechanics of the Earth's Atmosphere. Smithsonian Collections. Cleveland Abbe. Washington, D. C. 1910.

47. On the Representation of the Distribution of Atmospheric Pressure by Surfaces of Equal Pressure and by Isobars. Prof. D. Wm. von Bezold. Archives Neerlandaises des Sciences Exactes et Naturelles. Series 11. Tome VI. pp. 561-574. 1901. Also Mechanics of the Earth's Atmosphere. Smithsonian Collections. Cleveland Abbe. Washington, D. C. 1910.

48. Cyclone Memoirs. Part IV. W. L. Dallas. Meteorological Department of the Government of India. 8 vo. pp. 301-424. 16 Plates. Calcutta. 1891.

49. Passages of Low Areas over the Rocky Mountains. Cleveland Abbe. Monthly Weather Review. pp. 129-130. Washington, D. C. 1895.

50. Handbook of Cyclonic Storms in the Bay of Bengal. Sir John Elliott. Meteorological Department of the Government of India. 8 vo. pp. 212. 76 Plates. Calcutta. 1900-01.

51. West Indian Hurricanes. E. B. Garriott, Bulletin H. U. S. Weather Bureau. 4 to. pp. 68, Plates. Washington, D. C., 1900.

52. The Eye of the Storm. Sidney M. Ballou. American Meteorological Journal. Vol. 9. pp. 67-84 and 121-127. Boston, 1902.

53. The Cyclones of the Far East. Rev. Jose Algue, S. J. 1897. Second Revised Edition. Manila. 1904.

54. Apuntes Relatives a los Hurricanes de las Antillas on Septiembre y October de 1875 y 1876. R. P. Vinito Vines, S. J. Also Cyclonic Circulation and the Transitory Movements of West Indian Hurricanes. 8 vo. pp. 34. U. S. Weather Bureau, Washington, D. C.

55. The Law of Storms in the Eastern Seas. W. Doberck. Fourth Edition. Hong Kong. 1904.

56. Results of Some Empirical Researches as to the General Movements of the Atmospere. H. Hildebrand Hildebrandsson. U. S. Monthly Weather Review. June 1919. 47. pp. 374-389. Washington, D. C.

57. Les Bases de la Meteorologie Dynamique. H. Hildebrandsson and L. Teisserenc de Bort. 2 Vols. pp. 228 and 206. Paris. 1907.

58. Handbuch der Klimatologie. J. V. Hann. 8 vo. 3 Vols. pp. 394, 713, 426. Stuttgart. 1908-1911.

59. Hurricanes of the West Indies. Oliver L. Fassig. Bulletin X. pp. 28. Plates 42. U. S. Weather Bureau, Washington, D. C. 1913.

60. The Trade Winds in Puerto Rico. Oliver L. Fassiz. U. S. Monthly Weather Review. May, 1911. 39. Washington, D. C.

61. The Life History of Air Currents. Meteorological Office. London. 1906.

62. Forecasting Weather. W. N. Shaw. London. 1911.

63. Manual of Meteorology. Sir Napier Shaw. 4 Vols. London. 1926-1931.

64. Hurricanes and Revolving Storms. Mrs. E. V. Newnham. Geophysical Memoirs No. 19. Air Ministry. Meteorological Office. London. 1922.

65. On the Dynamics of Revolving Fluids. Lord Rayleigh. Proceedings of the Royal Society. London. 93a. pp. 148-154. 1917. Also Geophysical Memoirs No. 19. Air Ministry. Meteorological Office. London. 1922.

66. Types of Storms in the United States. E. H. Bowie and R. H. Weightman. U. S. Monthly Weather Review. Supplement I. pp. 37 Plates 114. Washington, D. C. 1915.

67. West Indies Hurricanes as Observed in Jamaica. Maxwell Hall. U. S. Monthly Weather Review, 45 pp. 587-588. December, 1917. Washington, D. C.

68. An Aerological Survey of the United States. Part I. Results of Observations by Means of Kites. Willis Ray Gregg. Monthly Weather Review. Supplement XX. Washington, D. C.

69. Tropical Cyclones. Iaac Monroe Cline. pp. XII and 301. Illustrated. New York. 1926.

70. Aeronautical Meteorology. Willis Ray Gregg. pp. 111 and 144. New York. 1925.

71. Relations of Changes in Storm Tides on the Coast of the Gulf of Mexico to the center and Movement of Hurricanes. Isaac Monroe Cline. U. S. Monthly Weather Review, March, 1920. 48. pp. 127-146. Washington, D. C.

72. Atlas of the Tracks of 620 Typhoons. 1893-1918. Rev. L. Froc. 4 to. Zi-Ka-Wei Observatory. Shanghaii. 1920.

73. Physics of the Air. Dr. William J. Humphreys, Second Editon. New York. 1929. Weather Bureau Edition. 1920.

74. The Law of the Geoidal Slope and Fallacies in Dynamic Meteorology. Dr. C. F. Marvin. U. S. Monthly Weather Review. 48. pp. 565-582. 1920. Washington, D. C.

75. The Energy of Cyclones, Discussion by Sir Oliver Lodge, Sir Napier Shaw, and others, in Nature, November to December 2, 1920. U. S. Weather Bureau, Monthly Weather Review. 49. pp. 3-5. January, 1921. Washington, D. C.

76. Tropical Cyclones, etc. Stephen S. Visher, U. S. Monthly Weather Review. 50. pp. 288-297 and 583-589. 1922. Washington, D. C.

77. Tropical Cyclones of the Pacific. Bernice P. Bishop Museum. Bulletin 20. Honolulu. 1925.

78. West Indian Hurricanes and Other Tropical Cyclones of the North Atlantic Ocean. Charles L. Mitchell. Supplement 24. U. S. Monthly Weather Review. Washington, D. C. 1924.

79. Trpoical Cyclones, by Isaac Monroe Cline. Reviewed by Dr. E. W. Barlow. Quarterly Journal of the Royal Meteorological Society. April, 1930. Vol. 56. pp. 203-205. London, England.

80. Tornado Studies for 1884. John P. Finley. Professional Papers of the U. S. Signal Service. No. XVI. Washington, D. C.

81. Tornadoes, What They Are, and How to Observe Them. John P. Finley. New York. 1887.

82. The Principles of Aerography. Alexander McAdie. Chicago and New York. 1917.

83. Storms of the Great Lakes. E. B. Garriott. Bulletin K. U. S. Weather Bureau, Washington, D. C. 1903.

84. International Cloud Observations. pp. 787. 162 Tables. 79 Charts. Annual Report, Chief U. S. Weather Bureau. 1898-99. Vol. II. Washington, D. C. 1900.

85. Publications of Harvard Univerity, Cambridge, Mass.

86. On the Structure of Moving Cyclones. J. Bjerknes. Geofysiske Publ. Vol. II. No. 3. Kristiana.

87. On the Dynamics of the Circular Vortex With Applications to the Atmosphere and Atmospheric Vortex and Wave Motions. V. Bjerknes. Geofysiske Publikationer. Vol. II. No. 4. Kristiana.

88. Life Cycle of Cyclones and the Polar Front Theory of Atmospheric Circulation. J. Bjerknes. Geofysiske Publikationer. Vol. III. No. 1 Kristiana.

89. On the Typhoons of the Far East. By Yosiki Hosiguti. Part 1. Memoirs of the Imperial Marine Observatory. Vol. II. No. 3. October 1926. Part II. Vol. III. No. 2 December 1927. Part III, IV and V, Vol. III. No. 3. June 1926. Kobe, Japan.

90. Dynamic Meteorology. H. C. Willett. National Research Council. Bulletin No. 79. Washington, D. C. 1931.

91. Thermodynamics Applied to Air Mass Analysis. C. G. Rossby. Massachusetts Institute of Technology. Meteorological Papers. Vol. I. No. 3. Boston, Mass. 1932.

92. American Air Mass Properties. H. C. Willett. Massachusetts Institute of Technology. Meteorological Papers. Vol. II. No. 2. Boston, Mass. 1933.

93. Reports on Norwegian Methods of Weather Analysis. F. W. Reichelderfer U. S. N. Washington, D. C. 1932.

94. West Indian Hurricanes. William H. Alexander. Bulletin No. 32. U. S. Weather Bureau. pp. 79. Washington, D. C. 1902.

95. Origin of Extra Tropical Cyclones. W. J. Humphreys. Read Before the American Meteorological Society. Chicago, Ill. June 19, 1933. U. S. Monthly Weather Review. Washington, D. C. 1933.

On the Occasion, or Incidental Causes, of Extra-tropical Cyclones. W. J. Humphreys, U. S. Monthly Weather Reviw. April, 1933. Vol. 61. pp. 112 and 12 charts. Washington, D. C. 1933.

96. Madagascar: Note Sur les Marees et Courants Cotiers Produits par les Cyclones Tropicaux. By Thomas, Chef du Service Meteorologique de Madagascar. Anales de la Commission pour l'Etude des Raz de Maree; No. 4. Paris. 1934.

97. Handbook of Cyclonic Storms in the Bay of Bengal by Sir John Elliott. Calcutta, 1900 and 1901.

98. A Popular Treatise on the Winds by William Ferrel, 1889. Second Edition, New York, 1911.

99. The West Indian Hurricane of September, 1919, in the Light of Soundings. R. Hanson Weightman, U. S. Monthly Weather Review October, 1919; 47, 717-720: Washington, D. C.

100. Birth and Death of Cyclones, by Sir Napier Shaw: Geophysical Memoirs, 19: Hurricanes and Tropical Storms by Newnham, 1922.

101. Mnual of Meteorology, Part IV. The Relation of the Wind to the Distribution of Barometric Pressure; by Sir Napier Shaw, Cambridge, 1919.

102. Hurricanes. Their Nature and History. By Ivan Ray Tannehill, B.S: Princeton University Press, 1938 .

103. Texas Monsoons. By Prof. Mark W. Harrington: Philosophical Society of Washington; Bulletin Vol. XII, 293-308.

Dr. Charles F. Brooks in The Bulletin of the American Meteorological Society, May, 1945, in a paper THE SOCIETY'S FIRST QUARTER OF A CENTURY commenting on my work in the Society makes reference to the material in Appendix A in the following:

"Isaac Monroe Cline, M.D., Ph.D., U. S. Weather Bureau, New Orleans, La., was our eighth president 1934 and 1935. Dr. Cline delivered as his presidential address "A Century of Progress in the Study of Cyclones," Pittsburgh, Pa., December 29, 1934. With eminent fairness, in a sweeping, comprehensive survey, Dr. Cline summarizes the contributions of the numerous investigators who have made significant additions to our knowledge of the nature of cyclones. He devoted especial attention to the tropical cyclone, on which he is an authority. The Society benefitted from Dr. Cline's presidency also through his keen interest and practical advice."

APPENDIX B

EDITORIALS

FROM THE MORNING TRIBUNE OF NEW ORLEANS, LA.

Wednesday, November 23, 1927.

HONOR EARNED

It was a graceful inspiration that impelled Vice-President Watkins, of the Southern Pacific Lines, to signalize the distinguished service of our old friend, Dr. Cline, by a testimonial dinner and a substantial memorial for the admiration of the Doctor's grandchildren.

Dr. Cline's cold and consistently correct calculation of prospective river stages, during the recent flood emergency, and the quiet pertinacity with which he published the results of it, were a blessing to many thousands in several states—both to those who were flooded and those who were not.

The accuracy of his conclusions occassioned relatively little note here at home, because we are accustomed to the Doctor being right about floods and hurricanes. His forecasts, however, aroused expressions of wondering admiration by cabinet officers from Washington, who were not so well acquainted with him.

He did the Southern Pacific a great service, as well as the commerce of New Orleans and our cross-country traffic in general a great service, in quietly laying the unsound hysteria that demanded the cutting of the S-P rail-embankments. But he did the Sunset Lines, relatively speaking, no more service than he did the humblest of the thousands who suffered by the flood, or thought themselves threatened by floods that did not arrive.

Dr. Cline is a man of more parts than one. Some of his contributions to hurricane-forecasting and general meteorology have been even more notable from a technical point of view, than his forecasting of flood stages. He is a man of healthy hobbies, aside from the generally dry, but sometimes very exciting, work at which he earns his living. He is an amateur of note in several other directions, as every busy man ought to be.

It is a great pleasure to join Mr. Watkins and his great railroad, in re-marking the Doctor's excellencies. For he is a modest man, and most of his fellow-citizens don't appreciate them.

* * *

FROM THE TIMES-PICAYUNE OF NEW ORLEANS, LA.
Thursday, January 12, 1928.

HONOR TO DR. CLINE

Out of the welter of worries and worse, due to the great flood of nineteen-twenty-seven, one of the unalloyed pleasures brought forth has been the wide and deep-felt recognition the event has brought to our own Dr. Isaac M. Cline, chief meteorologist in charge of the United States weather bureau. High officials at Washington, from their more distant viewpoint whence they were better able to see comprehensively the work of the scientist, were first to recognize the masterly manner in which Dr. Cline handled the emergency flood study and predictions and were quick to give credit to him for lives saved, suffering minimized and material loss avoided by the series of bulletins that will go down into meteorological history as exemplars of exactitude and timeliness.

Up to that hour the doctor's scientific light here in New Orleans had been dimmed by our familiarity with his long devotion to his important task. We, all of us, had ceased giving thought to the weather work in our acquired confidence that whatever there was to be done, be it ever so difficult or trying, Dr. Cline would do it, and do it thoroughly. Also our recognition of Dr. Cline, the forecaster and storm expert, was partly counteracted by our admiration for Dr. Cline, the lovable fellow citizen, the genial, cheerful, optimistic and splendidly honorable fellow Orleanian, for Dr. Cline, the lover of things beautiful, the connoisseur of art, the collector of many important works and donor to our museums of some of their most valuable historical pieces. "To know him is to love him," has never been spoken more truly than of this official who in manner and personality is as far apart as the poles from that which is often understood by the term bureaucrat. Be he within the precincts of his impressive government offices, or in the mellow light of his home walled with the works of the early American masters of portraiture, Dr. Cline is and ever has been the same warmly-welcoming and courteous gentleman.

Therefore, when President Coolidge, Secretary Hoover and others had shown the way to a more formal recognition, New Orleans' own people awoke to a realization that a collective manifestation of respect and admiration was due this devoted official's super-services to city, state and valley. As a preliminary, several weeks ago, the Southern Pacific railway, that had material reasons for gratitude for Dr. Cline's flood reports, paid special honors to the meteorologist. And now, as a suitable climax to a succession of honors, has come the presentation to the weather official of a loving cup bestowed jointly by four outstanding civic bodies of New Orleans, namely the Board of Trade, Association of Commerce, Cotton Exchange and New Orleans Steamship Association. Not only is this recognition deeply deserved, but also, we believe, Dr. Cline's sensitive and responsive nature is such that this is the kind of recognition that our splendid weather official, of all possible tributes, will value most highly.

* * *

FROM THE NEW YORK TIMES

Monday, January 23, 1928.

A WEATHER FORECASTER HONORED

It is rather the fashion when weather forecasts go wrong to belittle the work of the local and even the United States Bureau. One has heard owlish proposals to do away with the service. In his book on weather forecasting General A. W. Greely has said that predictions of what the weather would be outside of a limited zone could not be made accurately day in and day out. New atmospheric conditions are arising all the time, and the ink is hardly dry on a bulletin before it becomes necessary to modify it. Just now a "dead set" is being made against the head of the national bureau by an unofficial forecaster, and Mr. Marvin's usefulness might be ended if all that was said about him were true.

It is gratifying to know that servants of the people in the Weather Bureau are sometimes not without honor in their own country. There is the case of Dr. Isaac M. Cline of the New Orleans station, a gentleman gray in service, to whom the commercial organizations of New Orleans have just presented a silver amphora in recognition of the fact that his work during the Mississippi floods saved a great many lives and prevented the loss of property upon which no value could be set.

The Times-Picayune pays Dr. Cline a handsome but merited tribute when it says that "in manner and personality he is as far apart as the Poles from that which is often understood by the term bureaucrat;" that, on the contrary, he is "the lovable fellow-citizen, the genial, cheerful, optimistic and splendidly honorable Orleanian." Always regarded as a scientific and dependable forecaster, he rose to the emergency and, sacrificing his strength, gave himself unselfishly to humanity in the catastrophe of the floods.

* * *

My retirement from the weather service attracted widespread notice and complimentary editorials on my work appeared in many newspapers. The following editorial from the Dallas (Texas) Morning News, December 19, 1935, is representative of these comments.

* * *

DR. ISAAC CLINE RETIRES

After fifty-two years in the weather service, Dr. Isaac Cline of New Orleans is to retire. The Associated Press refers to him as "the smiling, genial climatological genius." Possibly the diction is a little faulty, but the alliteration is good, and those of our city who know the New Orleans meteorologist's brother, Dr. Joseph L. Cline, will understand. "Smiling, genial climatological genius" seems to be a trait of the Clan Cline.

Dr. Cline of New Orleans will open a shop in the old French quarter. This may surprise some people, but not those who know either Dr. Cline or even weather men as a class. Men who spend their lives looking at the sky are usually a little different. From the ranks of the meteorologists have come poets as well as scientists. Dr. Isaac Cline, himself, has attained considerable reputation as a writer.

Dr. Cline's fifty-two years of service have been one of the most intensively fought meteorological battle lines to be found anywhere in the world. The coast of the Gulf of Mexico has been the scene of as much spectacular weather performance as any other region outside the tropics, and in the tropics there usually is not such great life and property value to be fended against the fury of the elements. As the organizer of the Gulf Coast weather reporting service Dr. Isaac Cline has performed a service of incalculable benefit. He has earned greater reward than the Government can give. The best wishes of millions who have watched his daily reports for the Southern States are extended to him as he retires to the quiet of his little shop in the old French Quarter.

APPENDIX C.

JAPAN'S ALLY, THE TYPHOON, CAUGHT THE UNITED STATES NAVY UNPREPARED, OR OFF ITS GUARD. WHERE DOES THE RESPONSIBILITY FOR THIS DAMAGE REST?

Tropical cyclones (typhoons, Baguios, hurricanes, provincial names in different localities) are storms to be feared at all times and more especially when naval operations have to be carried on in regions where they are likely to be encountered. While the winds in the cyclone are destructive, studies of tropical cyclones in all parts of the world show that the most destructive force of the cyclone (typhoon) is the storm tide built up by and acting with the storm swells on coast lines. The storm swells move up from the rear of the cyclone through its right hand segment reaching coast lines well in advance of the cyclone and indicate the point towards which the center of the storm is advancing. As the cyclone approaches the coast the storm tide builds up in its right front and with the passage of the storm becomes the most destructive force of the cyclone. It is not the winds of 100 to 150 miles per hour that do the greatest damage and cause much loss of life but it is the storm swells and storm tides carried forward by these winds that leave destruction and death in their wake. A great deal has been learned during the last 25 years about the action of the winds and storm tides in tropical cyclones (see pages 182-188 this book). The strongest constructed ocean going craft are sometimes wrecked when frail craft not far distant ride safely to their goal. Accurate warnings of the cyclone should enable officers of ships to utilize available information and steer or shift their ships so as to avoid serious damage. History from earliest times gives incidents of the fury of the winds and the destruction caused by the storm tides in different parts of the world. Typhoons are of so frequent occurrence in the vicinity of Japan that they have been referred to as Japan's ally. During 1287-1291 Kublai Khan waged war on Japan and sent great fleets of war vessels which he was sure would be victorious. However, he did not reckon with the typhoons (cyclones) which virtually wiped out Kublai's fleets and forced him to abandon the conquest. Naval and maritime vessels have been greatly improved and strengthened since that time but the tropical cyclone remains a dreaded swirling monster of the atmosphere; these cyclones always inflict damage where they strike, whether on land or water. With the information now available relative to the destructive forces attending these storms and the part of the cyclone in which they occur it is surprising that ships do not more often maneuver so as to escape the area of greater danger.

When the United States was forced into World War II by the sneak attack on Pearl Harbor, December 7, 1941, by Japan it was certain that important naval battles would be fought in the regions about Japan and that naval vessels would encounter dangerous typhoons (tropical cyclones). Naval vessels operating in other waters might also encounter these dangerous storms. I was anxious to help in winning this war. I had 2,000 copies of my brochure A CENTURY OF PROGRESS IN THE STUDY OF CYCLONES, HURRICANES, TYPHOONS (reprinted as Appendix A in this book) for free distribution to Officers of the U. S. Army and the U. S. Navy. Several of the ranking officers of the navy and the army, acknowledged receipt of the publication with expressions of appreciation for the information contained therein. The following are the substance of their letters and show that they fully recognized the importance of the subject matter:

Admiral Chester W. Nimitz—Stated that he realized that such storms might be encountered in their operational areas and that he had passed the pamphlet to Fleet Meteorological Headquarters for study.

Lieut. General Robert G. Richardson, Jr., said that the weather has become of primary importance in making plans for military operations.

Vice Admiral C. A. Lockwood, Jr., Commander Submarine Force, Pacific Fleet—We have found many interesting and informative notes in this pamphlet.

Lieut. Col. Ray H. Martin (air pilot to Gen. Krueger)—Said that they had found the pamphlet very interesting and useful.

Lieut. Gen. F. M. Andrews—Said he had circulated the pamphlet through his staff of officers for study and reference.

The loss of naval vessels in the tropical cyclone which traveled northward in the Atlantic Ocean off the eastern coast of the United States in September, 1944, has been discussed on page 223 of this book. Damage to naval vessels in the region about the Philippines, and Japan was reported December 18, 1944, when a typhoon east of the Philippines, sunk three destroyers and damaged other vessels; more than 500 persons were reported dead or missing. June 5, 1945, a typhoon damaged 21 naval vessels; the cruiser Pittsburgh was so badly damaged that it had to go back for repairs. Notwithstanding the damage caused by the typhoon the main force of the fleet continued its operations successfully.

Okinawa was visited by a destructive typhoon, October 8 and 9, 1945. Early reports stated that 130 naval craft were beached or damaged. The Navy on October 14th identified three naval vessels that were sunk, three others damaged and thirty aground; twenty-eight navy men were reported killed, seventy missing; and 433 injured. The damage caused by the typhoons of December 18, 1944, and June 5, 1945 appears to have been inexcusably large when there is so much information available relative to the parts of the cyclone in which dangerous conditions are sure to be encountered. The beaching and damaging of so many naval craft in the Okinawa typhoon of October 8 and 9, 1945, leaves the impression that a high storm tide was an important contributing factor in causing the disaster. It appears that the naval craft were located on that part of the coast which would be traversed by the right hand segment of the typhoon. A rising storm tide which always precedes the arrival of the most dangerous part of the cyclone should have warned the naval officials far enough in advance to have enabled them to shift the craft to a point where the tide was not rising, a less dangerous location. The utilization of the storm tide in warning and protecting persons and shipping in threatened areas is fully explained in pages 158 and 159 of this book.

Some one was responsible for this disaster which in my opinion could have been in a great measure prevented. Statements given above show that the ranking admirals recognized the treachery of these storms and that they had passed the word down the line in a manner which is equivalent to—be prepared to meet them. They met them but apparently were not prepared.

On the opposite page is reproduced the front page of Weather Bureau Topics and Personnel, November, 1942, emphasizing the importance of confidence and promptness in meteorological work and citing my services for the information of all Weather Bureau personnel.

UNITED STATES DEPARTMENT OF COMMERCE

WEATHER BUREAU TOPICS AND PERSONNEL

NOVEMBER 1942

INFORMATION

WARTIME METEOROLOGICAL SERVICE

On October 20, Circular Letter No. 141-42 with enclosure referring to meteorological service for military representatives, was mailed to all Weather Bureau Offices. The contents of this Circular and the enclosure are important. Every employee is requested to read them.

The enclosure to the Circular Letter makes reference to the importance of confidence and promptness in meteorological service. Although these elements take on special importance in wartime, they are always essential elements in effective meteorological service. This fact is illustrated by a letter written to the Secretary of Agriculture by Herbert Hoover, then Secretary of Commerce, 15 years ago. It is worth quoting for information of all Weather Bureau personnel:

THE SECRETARY OF COMMERCE

Washington, D. C., July 5, 1927

Hon. WILLIAM M. JARDINE,
,Secretary of Agriculture,
Washington, D: C.

DEAR MR. SECRETARY:

There has been a service performed during the flood in Louisiana * * * which should be made a matter of illuminated record in your archives.

Dr. Cline made a series of day to day estimates of the flood and its progress for over 2 months of its movement from Cairo to the Gulf. They were based upon a wealth of technical understanding and experience that was almost uncanny. We came to rely upon them and to build the whole organization and direction of our rescue operation upon them. With the advance information which he was able to give us on; the action of the flood we were able to interpret it into a saving of thousands of human lives. Moreover the issue of this material to the public by Dr. Cline in such terms as to give confidence and understanding resulted in the saving of great life and property without any action on our part. His office was in service day and night with no demand that we could make that was not more than answered. Without the accurate information which he supplied I believe that New Orleans might have gone under water.

"It has been much more than the mere routine interpretation of technical data. It required judgment and discretion, which amounted to genius. He has been an honor to the Weather Bureau and the whole Department.

Faithfully yours,

(Signed) HERBERT HOOVER